Rüdeger Baumann

BASIC Game Plans
Computer Games and Puzzles Programmed in BASIC

Thomas S. Hansen, *Editor and Translator*
Donald Kahn, *Technical Editor*

Birkhäuser
Boston · Basel

Rüdeger Baumann
Lilo-Gödenstrasse 16
2120 Lüneburg
Federal Republic of Germany

Thomas S. Hansen
Department of German
Wellesley College
Wellesley, MA 02181
U.S.A.

Donald Kahn
P.O. Box 10
Colrain, MA 01340
U.S.A.

Library of Congress Cataloging-in-Publication Data
Baumann, R. (Rüdeger)
   BASIC game plans.
   Bibliography: p.
· 1. Computer games.   2. Basic (Computer program
language) I. Hansen, T. (Thomas)   II. Title.
GV1469.2.B38   1988   794.8'2   83-27551
ISBN 0-8176-3366-9

CIP-Titelaufnahme der Deutschen Bibliothek
Baumann, Ruedeger:
BASIC game plans : computer games and puzzles programmed
in BASIC / by Ruedeger Baumann. Edited and translated by
Thomas S. Hansen. Techn. ed.: Donald Kahn. - Basel ; Boston:
Birkhäuser, 1988
   ISBN 3-7643-3366-9 (Basel) brosch.
   ISBN 0-8176-3366-9 (Boston) brosch.
NE: Hansen, Thomas S. [Bearb.]

Printed on acid-free paper.

Originally published under the title *Computerspiele und Knobeleien programmiert in BASIC* by Vogel-
Verlag, Würzburg (German Federal Republic). Copyright 1982 by Vogel-Verlag, Würzburg. English
translation based on second edition (1983). Copyright of the English translation 1989 by Birkhäuser
Boston Inc., Cambridge, Massachusetts, U.S.A.

ISBN 0-8176-3366-9
ISBN 3-7643-3366-9

Text provided in camera-ready form by the editor.
Printed and bound by Edwards Brothers Incorporated, Ann Arbor, Michigan.
Printed in the U.S.A.

9 8 7 6 5 4 3 2 1

FOREWORD

> The computer is a toy tossed to us by
> Nature for diversion and comfort in the
> darkness.
>
> d'Alembert

> I hate everything which merely instructs
> without stimulating me or increasing my own
> activity.
>
> Goethe

Let's try to eliminate some misconceptions from the outset: this book is <u>not</u> a collection of game recipes assembled in the form of finished programs which only have to be typed into the computer and then played. Far from it. The object is to challenge readers to activate their own creativity in using computer games. The game concept is designed to develop into game strategy and this then should form the basis of computer programming. Programming computers is in itself the game. Or, to put it another way, readers can learn programming while playing.

No previous knowledge of programming is assumed of readers and users of this book -- only the willingness to accept new ideas and improve upon them independently. While all the programs in this book have been run and tested, some are intentionally imperfect. They await the finishing touches from you, the reader. The additional brainteasers at the end of the chapters (or, occasionally, after a section within a chapter) are therefore designed to inspire your imagination and encourage your independence.

The material is drawn from numerous sources. Martin Gardner deserves particular mention as his books and his columns in Scientific American are a great treasure trove of game ideas. Furthermore, the journals <u>Creative Computing</u> and <u>Jeux et Strategies</u> deserve mention. Gardner's influence can be felt in their pages although he himself also draws upon other writers in this field. Very often it is impossible to determine the original inventor of a computer game.

Game playing is as old as the human race; all peoples and cultures know and promote the activity. In our times game playing has received new impetus through the so-called "Third Industrial Revolution," on the one hand because our leisure time has been increased, and on the other, because the computer has provided us with a partner and master player of remarkable talent and versatility. It is fascinating to be able to instruct a machine so that it can be taken seriously as a game opponent. The computer has such universal talent that every game -- whether a game of survival or mere entertainment -- can be simulated on the machine or played against it.

Some people think that games lose their appeal once the theories behind them are known. For the lover of computers, however, this is just where things get interesting. Playing may be fun but programming the game is more so. After all, one can't play the game until it has been thoroughly understood. Game playing is an activity very distinct from work in that it is practiced for its own sake, or rather, for the sake of the continuous pleasure it provides -- even (or maybe especially) in our technologized world. The computer -- so often maligned as a force opposed to human values -- is thus contributing to the aesthetic education of mankind.

R. Baumann   Lueneburg (West Germany)

If you would like a 5 1/4 inch, Commodore compatible diskette for the programs in this book, please send $5.00 to:

Donald Kahn
P.O. Box 80
Colrain, MA 01340

TABLE OF CONTENTS

A WORD TO THE USER

The computer programs in this book were originally conceived for the Commodore PET. Nowadays few readers still have access to this famous machine that helped to revolutionize the world of personal computing. Its popular successors that are more frequently found in private use include the Commodore 64, the Apple II, and the IBM PC (and compatible machines). This book and its programs have been revised for users of such modern hardware. These three quite different computers still have much in common. For example, they share common languages, of which BASIC is one, although there are differences between the various versions of BASIC when run on different machines. Furthermore, each machine is different internally, so the use of POKE (or PEEK) can be slightly different. All such details are explained in full below so that you will be able to write and run the BASIC programs in this book on your own machine.

In keeping with the spirit of the rest of the book, instructions and technical information are provided, but the implementation is, for the most part, left to the reader.

PET BASIC

There are only a few differences between Commodore BASIC and Apple and IBM BASICs which affect the programs in this book. The GET A$ command is replaced on the IBM PC with A$=INKEY$. Also, IBM PC BASIC does not allow you to omit GO TO following an IF...THEN statement. The Commodore variable TI reads a clock in the Commodore; the IBM's clock is read with TIMER, but the Apple II requires an add-on card for this function. Reseeding the Commodore random number generator is accomplished most randomly with RND(-TI). Use RANDOMIZE TIMER on an IBM PC. Those of you with Apples may need to use your imagination to most randomly reseed.

PET Graphics

The computer programs in this book use two methods of display, PRINTing and POKEing. (All programs may be assumed to be in text mode.) The table below indicates the meaning of PRINTing the eight CHR$ codes which represent screen control commands, as well as the closest equivalent commands for the Apple and the IBM.

Function	Commodore PRINT	AppleSoft II command	IBM BASIC command
Clearing the screen	CHR$(147)	HOME	CLS
Cursor to home (upper left)	CHR$(19)	VTAB 1: HTAB 1	LOCATE 1,1 or PRINT CHR$(11)
Right cursor	CHR$(29)	HTAB or PRINT " "	PRINT CHR$(28)
Left cursor	CHR$(157)	HTAB	PRINT CHR$(29)
Down cursor	CHR$(17)	VTAB	PRINT CHR$(31)
Up cursor	CHR$(145)	VTAB	PRINT CHR$(30)
Reverse on	CHR$(18)	INVERSE	COLOR x,y
Reverse off	CHR$(146)	NORMAL	COLOR y,x

Also, moving the cursor around the screen may be accomplished by POKEing. In the Apple II, address 36 contains the cursor's horizontal position, and address 37 contains the cursor's vertical position.

All three computers allow POKEing as a means of display. The memory addresses in the PET which correspond to the 1000 print positions on the screen (1000 = 25 rows * 40 columns) are 32768-33767. To display a 4 in the center of the screen, one would POKE 33267,52. (33267 = 32768 + 499, and 52 is the screen code for a 4.)

Similarly, the Commodore 64 has 1000 screen locations. They reside in memory in addresses 1024-2023. The Apple II has 24 lines of 40 positions, for a total of 960, and they live in addresses 1024-2047; the extra 64 addresses are for peripheral boards. Unfortunately, these extra addresses are scattered throughout the display memory. After 120 locations, ther are 8 card (board) locations; this pattern is repeated eight times. The IBM PC has 24 rows of either 40 or 80 columns. With a graphics board and a monochrome display, the statement SCREEN 0:DEF SEG=&HB000 will set screen memory in even locations 0-2000.

For the Apple II and the IBM PC, the number to POKE is the ASCII code for the character desired. Commodore screen codes may be different from their ASCII values. For example, the letter A has a screen code of 1 and an ASCII code of 65. With an Apple II, ASCII value 65 represents a flashing A; a constant A is 193. On the IBM PC, A is 65.

Users with an Apple or an IBM may be using peripheral graphics boards. Refer to your instruction manuals to determine how screen memory is different with the use of these boards.

The PET has a separate numeric key pad which is used in several of the programs in chapter 3. If your computer does not have a numeric key pad, use the regular keyboard (choose four keys in a cross pattern) or PEEK a joystick.

PET Memory

In addition to screen memory, some programs in this book refer to a few other memory addresses. Program/example 2.6 uses the keyboard buffer and the buffer counter.

	PET	Commodore 64	IBM PC (DEF SEG=0)
keyboard buffer	623-632	631-640	1054-1085
buffer counter	158	198	1050

The Apple II has a character input buffer in memory locations 512-767, and the last character typed goes to address 49152. POKEing or PEEKing 49168 clears 49152.

The Mini-chess program uses address 59468 to switch between upper case (12) and lower case (14). With a Commodore 64, use 53272 (21,23). With an IBM PC, use (DEF SEG=0) POKE 1047,PEEK(1047) AND 191 for lower case, and POKE 1047,PEEK(1047) OR 64 for upper case. The bare (no boards) Apple II has no lower case, but by the time you get to chapter 6, you will be knowledgeable enough to overcome that minor inconvenience.

PET Sound

Commodore grestly improved the sound capabilities of its computers with the Commodore 64's SID chip. In the chapter 3 discussion of sound, three addresses are used:

```
59467   (tone on/off)
59464   (pitch)
59466   (timbre)
```

The Commodore 64 SID chip controls three voices, the PET only one. The 64 also has addresses which control filtering (high pass, low pass, or bandpass) and modulation, as well as two addresses for the pitch. The addresses for voice 1, corresponding to those above, are:

```
54296   (volume)
54272,54273   (pitch)
54276   (envelope)
```

IBM PC BASIC has music commands, eliminating the need to POKE. SOUND f,d (pitch f, duration d) may be useful. The PLAY command, for music composition, is beyond the scope of this book.

A beep may be generated on the Apple II by PEEKing (-16336), but a machine language program or an add-on board is necessary to actually produce music.

* * * * *

In closing, let me say a few things. First, a list of helpful reference books follows. Second, all of the programs in this book were tested on a Commodore 64; if your program does not run, it is probably due to a typing error. Finally, if you enjoy games, math, or programming, you will enjoy this book; I know I did.

Donald Kahn, technical editor
August, 1987

Apple Programming Secrets, by Wayne Dyrness
AppleSoft II, Apple Computer Inc.
Apple II Reference Manual, Apple Computer Inc.
What's Where in the Apple?, by William F. Luebbert

Commodore 64 Programmer's Reference Guide, Commodore Business Machines, Inc.

Handbook of BASIC for the IBM PC, by David I. Schneider

BASIC Program Conversions, by Editors of Computer Skill Builders; Bill Crider, Managing Editor

INTRODUCTION

The interest in games is immense and growing steadily. Human imagination on the subject seems inexhaustible. So as not to drown in the flood of material surrounding us, we have to restrict ourselves to a small portion of what is available. Fortunately, when it comes to computer games, this restriction still offers wide choices. By grouping certain games together we get an overview of the subject. Here, then, are some of the main themes and principles behind the organization of this book.

1. First of all, certain games can be arranged according to the computer's role in the process:

   a) The computer can be the master player. It can monitor the rules as well as the progress of the game. It evaluates the players' moves and judges who is the winner and who is the loser.

   b) The computer can also be a game partner or an opponent. If this is its role we must then give it certain intelligence (unless we are playing games of pure chance). This is probably the most interesting -- and also the most challenging -- of programming tasks. If you empower the computer with its own strategy, you first have to have thoroughly understood and analyzed the game.

2. A further method of grouping games is by number of players:

   a) Games involving one player are the famous brain teasers. These are puzzles that demand patience or concentration. Here the computer is given the role of master player and wizard. Thanks to its recombinational ability it can search rapidly for all the possible solutions to these puzzles (see chap. 8 and 9).

   b) Games with two players are the subject of chapters 4 through 7. In this situation the computer and the game player can exchange roles and try different strategies against each other.

c) Games with several players are less common. Nonetheless, the reader is often asked to expand a game to include more than two players, and to include the computer as expert or simply as one of the regular players.

3. The third group of games includes those in which chance is a factor.

a) If we exclude chance so that the player's decisions are the only factor determining the game's progress, then we have a game of pure strategy (see chapter 6).

b) In games of pure luck, however, the player cannot affect the progress of the game. These are the games that demonstrate the operation of chance. Have patience with the computer if it can't recreate the smoke-filled atmosphere of the gambling casino. Until such technology is available on a personal computer, your imagination will have to suffice (chapter 5).

c) In games of mixed strategy the player builds an element of chance into his strategy. Games of this type are fascinating but exceedingly complex.

Up to this point we have not been categorizing games according to the equipment required to play them, i.e., board, dice, cards, pen or pencil. These props are hard to represent on the screen without computer graphics, which is the subject of chapter 3. But on the whole, games can be played graphically on a videoscreen.

If we organize the games covered in this book according to the player's role, we get the following list:

-- Games of Searching and Guessing (chapter 4).
-- Concentration and Memory Games (chapter 2).
-- Games of Skill and Reaction Time (chapter 3).
-- Mind Games

This organization may not seem strictly consistent. In games involving searching and guessing, for example, you can't avoid using your mind.

One more word about BASIC, the programming language used here. The justifiable charge made against this language is that it can lead users to engage in wild, unsystematic programming resulting in the dreaded "spaghetti programming." Salient examples of this are certain programs to be found in the current literature. There are countless titles on the market that contain some thoroughly unreadable and unintelligible programs. A major objective of this volume, however, is to prove that even beginners can write well-structured and intelligible programs in BASIC. Then, there will always be the speed fanatics who turn up their noses at this programming language and cry "too slow!". Computer users without great storage capacity may find the length of many programs wasteful. However, once you have grasped the mechanics of a program, it is easy to create faster, shorter versions.

Finally, readers with criticisms and creative suggestions that could improve this book are urged to contact Ruedeger Baumann: Lilo-Goeden-Strasse 16, 2120 Lueneburg, West Germany; or Thomas S. Hansen: Founders Hall 117, Wellesley College, Wellesley, Mass. 02181 (USA)

CHAPTER I
RULES OF THE GAME:
INTRODUCTION TO PLAYING WITH AND AGAINST COMPUTERS

The games in this section are (with the exception of number 4) simple in two senses of the word: they are both easy to play as well as easy to program. They are meant to stimulate you, the user, to become familiar with the computer. In this first little game the computer will seem to have telepathic powers. It is actually an old trick based on the values of different coins.

Example 1.1 WHICH HAND?

The computer asks the player to hold a dime in one fist and a penny in the other. Then it asks the holder to multiply the value of the coin in the right hand by 8 (or by any other even number) and to multiply the value of the coin in the left hand by 5 (or by any other odd number). The player then adds the numbers together and tells the computer whether the result is odd or even. Then the computer tells the player which hand holds the dime and which one holds the penny.

Dialogue between computer and user might look like this:

```
Computer: Is the sum of the two numbers you added
odd or even?
Player: Even.
Computer: The dime is in your left hand and the penny
is in your right.
```

Before explaining the technique behind the trick, let's write the program for it. The process behind the program can be described as follows:

```
Display: Rules of Play and Question
              (* of the computer *)
Enter: Answer (* of the player *)
IF Answer = "EVEN"  THEN
      Hand with dime:=left
      Hand with penny:=right
OTHERWISE
      Hand with dime:=right
      Hand with penny:=left
END IF
Display:
    "The dime is in "; hand with dime;
    "The penny is in "; hand with penny
```

How do you explain the trick? There are two possibilities:
1. 1st case: dime right, penny left.

Value right: 8*10, general: 2m*10   (even number)

Value left: 5*1, general: (2n+1)*1 (odd number)

Total: 85,   general: 2(10m+n)+1 (odd number)

2. 2nd case: dime left, penny right.

Value right: 8*1, general: 2m*1   (even number)

Value left: 5*10, general: (2n+1)*10 (even number)

Total: 58,   general: 2(m+10n+5) (even number).

The explanation for the trick is the rule: "even + even = even" and "even + odd = odd."

Here is this process translated into the programming language.

```
100 : PRINT CHR$(147) : REM CLEAR THE SCREEN
110 : PRINT "            WHICH HAND?
111 : PRINT "            _____
112 :
120 REM THE COMPUTER FUNCTIONS AS MAGICIAN.  IT SEEMS TO
121 REM BE ABLE TO READ MINDS THANKS TO A SIMPLE ARITHMETIC
122 REM RELATION OF EVEN NUMBERS.
123 :
200 REM ===THE RULES OF THE GAME===========================
201 :
210 : PRINT : PRINT
220 : PRINT "PICK UP A DIME IN ONE HAND AND A PENNY IN THE
```

```
230 : PRINT "OTHER. I'LL GUESS WHICH HAND HOLDS WHICH COIN
240 : PRINT "IF YOU ANSWER A COUPLE OF QUESTIONS FOR ME.
245 : PRINT "MULTIPLY THE VALUE OF THE COIN IN YOUR RIGHT
250 : PRINT "HAND BY 8 (OR WITH ANY OTHER EVEN NUMBER) AND
255 : PRINT "MULTIPLY THE VALUE OF THE COIN IN YOUR LEFT
260 : PRINT "HAND BY 5 (OR ANY OTHER ODD NUMBER).
290 :
300 REM=======QUESTION AND ANSWER===========================
301 :
310 : PRINT
320 : PRINT "IS THE SUM OF THE TWO NUMBERS YOU ADDED ODD
330 : PRINT "OR EVEN?
335 : PRINT "PLEASE ANSWER 'ODD' OR 'EVEN.'
340 : PRINT
350 : INPUT "YOUR ANSWER   "; A$
360 :
370 : IF A$ = "EVEN"    THEN 400
380 : IF A$ = "ODD"     THEN 450
390 : PRINT "WRONG INPUT!": GOTO 335
395 :
400 : REM THE SUM IS EVEN.
405 :
410 : LET D$ = "LEFT"    : REM DIME IN LEFT HAND.
420 : LET P$ = "RIGHT"   : REM PENNY IN RIGHT HAND.
430 : GOTO 500
440 :
450 : REM THE SUM IS ODD.
455 :
460 : LET P$ = "LEFT"    : REM PENNY IN LEFT HAND.
470 : LET D$ = "RIGHT"   : REM DIME IN RIGHT HAND.
490 :
500 REM ======THE COMPUTER'S ANSWER=========================
501 :
510 : PRINT : PRINT
520 :
530 : PRINT "THE DIME IS "; D$;", THE PENNY ";P$"."
540 :
590 : END
```

Explanation of the program WHICH HAND?

1. Line 100 clears the screen.

2. The term REM means "remark" and introduces a comment telling the computer to ignore what follows on the same line. If you are entering the program by hand and do not want to spend a lot of time typing, simply omit lines 120-122, 200, 300, 400, etc. You must then change the skip commands (GOTO), which find these lines (line 430 must read "GOTO 510"). Of course, you don't need to enter lines con-

3

taining no commands, such as line 270. The leading colons may also be omitted.

3. Line 350: The variable A$ (this notation is read "A string") will receive your answer. The dollar sign ($) marks a string variable, which is an ordered collection of characters or numbers. The command INPUT tells you to enter your answer.

4. In line 370 there is an IF command:

```
IF A$ = "EVEN"   THEN 400
```

This means:

```
IF A$ = "EVEN" THEN
    proceed with program line 400
OTHERWISE
    proceed with the following line (380)
END IF
```

If the player has not answered either "even" or "odd" then line 390 causes a return to line 335 and repeats the request for an answer.

And now for something more useful where both the programming and the game playing are on a slightly higher level but no less entertaining.

Example 1.2 ADDITION WARM-UPS
The computer poses an addition problem and the user enters the solution. The computer then answers with "correct" or "incorrect" and proceeds to a new problem if the answer is correct. The number of problems can be chosen in advance.

Here is the dialogue between player and computer:

ADDITION WARM-UPS

```
How many problems do you want to do? 2
How much is 28+13? 41
Correct!
How much is 39+15? 64
Sorry, that's wrong. Try again.
How much is 39+15? 54
Correct!
Want to try another problem? (y/n) ?  n
It's been a pleasure. Good-bye.
```

The process behind the computer's activity is as follows:

ADDITION WARM-UPS

```
Enter: number (* of problems desired *)
FOR i equals 1 TO number REPEAT
        Choose two random numbers: z₁, z₂
        between 1 and 100.
        Question: How much is z₁+z₂
        Enter: total
        IF z₁+z₂ = total   THEN
                response: Correct!
        OTHERWISE
                response: Incorrect!
        END IF
END REPEAT
```

Here we don't make the computer repeat the question when the answer is incorrect, although the program will do so. The BASIC program is as follows:

```
100 : PRINT CHR$(147)
110 : PRINT "               ADDITION WARM-UPS
111 : PRINT "            .    _____
112 :
120 : REM THE OBJECT IS TO IMPROVE YOUR SKILL AT ADDING
121 : REM NUMBERS.
129 :
130 : PRINT
140 : INPUT "HOW MANY PROBLEMS DO YOU WANT "; N
190 :
```

5

```
200 : FOR I = 1 TO N  : REM MAIN LOOP =======================
201 :
210 :   REM ------CHOOSING TWO RANDOM NUMBERS----------------
220 :
230 :      LET Z1 = INT(99*RND(1))+1
240 :      LET Z2 = INT(99*RND(1))+1
250 :
260 :   REM ------- QUESTION AND ANSWER--------------------
270 :
280 :    PRINT
290 :    PRINT "HOW MUCH IS";Z1;"+";Z2
300 :    INPUT S
310 :    PRINT
320 :    IF S = Z1+Z2 THEN PRINT "CORRECT!" : GOTO 360
330 :    PRINT "INCORRECT";
340 :    PRINT "TRY AGAIN.": GOTO 280
350 :
360 : NEXT I  : REM END OF THE MAIN LOOP====================
390 :
400 : REM ----- EXIT OR REPEAT --------------------------
401 :
410 : PRINT  : PRINT  : PRINT
420 : INPUT "WANT TO TRY ANOTHER PROBLEM (Y/N) "; A$
430 : IF  A$ = "Y" THEN RUN
440 : PRINT
450 : PRINT "IT'S BEEN A PLEASURE. GOOD-BYE."
460 : PRINT : PRINT
490 : END
```

Explanation:

1. A new element in this program is the use of a "loop" to repeat instructions. Here is a so-called "counting loop":

```
200   FOR I = 1 TO N
      <commands>
360   NEXT I
```

The variable I counts from 1 to N. The commands are carried out at every loop, i.e., executed exactly N times (if N is greater than or equal to 1).

2. In order to have the computer keep posing new problems, we insert the function RND(1) in lines 230 and 240 which gives us random numbers between 0 and 1. If the random numbers are to be whole numbers between A and B, then we must enter:

$$LET\ Z=INT((B-A+1)*RND(1))+A$$

For A=1 and B=99 the result is:

$$LET\ Z=INT(99*RND(1))+1.$$

3. Line 320 tests the program to see if the user's answer is correct. If this is the case, then it continues with line 360 (NEXT I), which means that it raises the counting variable I by a factor of one and supplies the next problem, assuming N has not been exceeded.

4. Line 204 asks for a possible repetition. If the answer was "Y" the program begins again from the beginning (command RUN). Otherwise it says good-bye.

****

Let's take a simple question and answer game for our third example:

Example 1.3 TOO HIGH -- TOO LOW (Number guessing)
The object of this game is to guess a number between 1 and let's say, 14, chosen at random by the computer. After each guess the computer adds a hint "too high" or "too low." When the correct number is guessed, the computer reminds the player how many tries it took.

A dialogue can look like this:

```
            TOO HIGH - TOO LOW
I am thinking of a number between 1 and 14.
What is the number? 7
7 is too high.
What is the number? 4
4 is too low.
What is the number? 5
Bulls-eye!
That took you 3 tries.
```

Just to keep the guessing from getting too simple, the computer randomly sets the upper limit of the number region. In the previous dialogue the computer had chosen 14 as the upper limit.

Here is a description of the process:

```
                   TOO HIGH - TOO LOW
 Upper limit:=random number between 1 and 1000
 Number chosen:=random number between 1 and upper limit
 Display: I am thinking of a number between 1 and "upper
          limit."
 Number of tries:=0
 REPEAT
   REPEAT
     Display: "What is the number?
     Enter: player guesses number
     Number of tries:= number of tries + 1
        IF number guessed > number chosen THEN
           Display: Number guessed "is too low"
        BUT IF number guessed is < number chosen THEN
           Display: Number guessed "is too low"
        OTHERWISE
           Display: Bull's eye!
        END IF
   UNTIL number guessed=number chosen
   END REPEAT
     Display: "One more time?"
     Enter: answer
 UNTIL  answer does not equal "yes"
 END REPEAT
 Display: "Good-bye."
```

Here is the BASIC program for the game:

```
100 : PRINT CHR$(147)
110 : PRINT "            GUESS THE NUMBER
111 : PRINT "            _____
112 :
120 REM THE OBJECT IS TO GUESS A NUMBER THE COMPUTER KNOWS.
190 :
200 REM ====SETTING THE BEGINNING VALUES====================
201 :
210 : LET UL = INT(1000*RND(1))+1: REM UPPER LIMIT OF THE
215 : REM NUMBERS
220 : LET N = INT(UL*RND(1))+1 : REM NUMBER REMEMBERED
230 : PRINT
240 : PRINT "I AM THINKING OF A NUMBER BETWEEN 1 AND" UL
```

```
250 :
260 : LET T = 0       : REM NUMBER OF TRIES
290 :
300 : REM===========GUESSING PROCESS=====================
310 :
310 : PRINT
320 : INPUT "WHAT IS THE NUMBER"; G
330 : LET T = T+1            :REM ONE MORE TRY
340 : IF G > N THEN PRINT G;"IS TOO HIGH": GOTO 310
350 : IF G < N THEN PRINT G;"IS TOO LOW": GOTO 310
360 : PRINT  : PRINT : PRINT  "CORRECT!";
370 : PRINT  " YOU NEEDED ";T;" TRIES."
390 :
400 REM===========REPEAT OR EXIT=====================
410 :  PRINT  :  PRINT  :  PRINT
401 :
420 : PRINT "ONE MORE TIME (Y/N)?";
430 : GET A$: IF A$ ="" THEN 430
440 : IF A$ = "Y" THEN PRINT CHR$(147): GOTO 200
450 : PRINT : PRINT:  PRINT: "GOOD-BYE."
490 : END
```

Explanation:

1. Line 210 sets the upper limit of the numerical region (1 - 1000) with the symbol UL. The rest of the commands are already familiar.

2. Line 430, however, contains a new command: the command GET A$ makes it possible to enter exactly one symbol from the keyboard into the variable A$ without holding up the program (as with INPUT). With the command

```
IF A$ = "" THEN 430
```

the computer is asked whether a character was entered (into A$). If no character was entered, A$="" (an empty character string), and the program remains within the loop of 430. It will only leave this and continue with the program when a character has been entered into A$. Do not confuse the empty character string "" with " " (with space between the quotation marks). The following example introduces the concept of Game Strategy which is basic to game playing.

9

Example 1.4 TARGET 100

Beginning with a randomly chosen number between 1 and 30, the players (one of whom is the computer) add integers between 1 and 10. The player who reaches 100 is the winner.

A typical dialogue for this game looks like this:

```
Beginning number: 14
How much do you want to add? 6
New total: 20
Adding  3
New total: 23
How much do you want to add? 5
New total: 28
(........................)
Adding 4
New total: 89
How much do you want to add? 8
New Total: 97
Adding  3
New total: 100
I win!
Another round? (y/n)   n
Thank you. It was a pleasure.
```

Now we have to program the computer so that it can carry on this intelligent dialogue and so that it also stands a chance to win.

Each round of the game consists of a series of game positions, which are simply the new total reached each time (in the game played above these were 14, 20, 23, 28, ..., 89, 97, 100). These totals mark the game's progress by creating new "game positions." Now we have to ask why the human player lost against the computer?

Look at the last total reached before 100. It was 97 and was a clear winning position because the next move (computer adds 3) won the game. Each of the numbers 90, 91, ..., 99 would have brought a winning positon for the player whose turn it was. 89, however, would have been a losing position because each following move would have resulted in a winning position for the opponent. Similarly, we can identify 78, 67, 56, 45, 34, 23, 12, 1 as losing positions and all other numbers as winning positions.

We shall program the computer so that it always plays toward these losing numbers (or losing positions) which it then hands to the player to make the best of. However, when the player hands a losing number to the computer, it will always respond by adding 1. The computer's behaviour is an example of "game strategy": for every situation, the computer has the correct counter-move ready.

The BASIC program is printed below. It is divided into the sections:

> Selection of starting number
> Player's move
> New Total
> Computer's move
> New Total
> Another round?

In lines 520-540 the computer runs through a small loop to find the next winning position (its name for the losing positions of the opponent). This winning position is the smallest of the numbers G which is also larger than or equal to game position S. RESTORE causes the computer to read the data line 810 from the beginning with each READ.

If G = S then the computer gives the opponent the edge (line 550) by making the least move possible, 1. Otherwise it adds to S a number that will create a losing position for the opponent.

```
100 : PRINT CHR$(147)
110 : PRINT "          TARGET 100
111 : PRINT "          _____
112 :
120 : REM THIS IS AN EASY GAME SIMILAR TO NIM.
121 :
130 : PRINT "IN EVERY MOVE ADD A NUMBER BETWEEN  1 AND 10
140 : PRINT "TO THE PREVIOUS TOTAL.
150 : PRINT "THE FIRST PLAYER TO REACH 100 IS THE WINNER.
200 REM ========CHOOSING THE STARTING NUMBER===============
201 :
```

```
210 : LET S = INT(30*RND(1)+1)
230 : PRINT   "STARTING NUMBER: "; S
290 :
300 REM ====PLAYER'S MOVE=====================================
301 :
320 : INPUT  "HOW MUCH ARE YOU ADDING?"; N
330 : IF N<1 OR N>10 THEN PRINT "WRONG ANSWER.": GOTO 320
340 : LET S = S+N
350 : PRINT "NEW TOTAL:"; S
390 :
400 REM=======HAS NUMBER BEEN REACHED?=====================
401 :
410 : IF S >= 100 THEN PRINT: PRINT "YOU WIN.": GOTO 700
490 :
500 REM======COMPUTER'S MOVE=================================
501 :
510 :   REM ------ DERIVING THE NEXT WINNING NUMBER----------
515 :
520 :     RESTORE
530 :
535 :     READ G : REM READ WINNING NUMBER
540 :     IF G < S THEN GOTO 535
544 :
545 :   REM -----COMPUTER'S MOVE-----------------------------
546 :
550 :     IF G = S THEN LET Z = 1 : GOTO 570 : REM COMPUTER
555 :       REM GIVES OPPONENT EDGE
560 :     LET Z = G-S : REM DIFFERENCE FROM NEXT WINNING
565 :       REM NUMBER
570 :     PRINT : PRINT  "I AM ADDING " ; Z
580 :     LET S = S+Z
590 :     PRINT "NEW TOTAL: "; S
599 :
600 REM=============HAS NUMBER BEEN REACHED?================
601 :
610 : IF S < 100 THEN GOTO 300
620 : PRINT : PRINT  "SORRY, YOU HAVE LOST."
690 :
700 REM ====ANOTHER ROUND?=================================
701 :
710 : PRINT: PRINT  "ANOTHER ROUND? (Y/N)"
720 : GET  A$:  IF A$ = "" THEN GOTO 720
730 : IF A$ = "Y" THEN PRINT CHR$(147): GOTO 200
740 : PRINT
750 : PRINT "THANKS, IT WAS A PLEASURE."
790 :
800 REM=============WINNING NUMBERS=====================
801 :
810 : DATA 1, 12, 23, 34, 45, 56, 67, 78, 89, 100
880 :
890 : END
```

All the games we have practiced up to now have been constructed around a dialogue between computer and player. To
conclude this introductory section let's work on a game
meant for several players. Introduce your friends to this
one some evening.

Example 1.5 MATCH THE PARTNERS
The computer asks your guests, both male and female, certain questions. Then it displays a set of numbers for each
two persons (one male, one female). These numbers match up
"couples" based on their individual answers.

In order to present this rather long program clearly, we
have divided it into main and subroutines. The main program
will call the following subroutines:

        1. Initialize

        2. Enter name

        3. Question and Answer

        4. Evaluation

        5. Exit

Step 1 sets the dimensions for tables of names and
answers. In other words, it tells the computer to prepare
storage space. Then it enters the questions. If you don't
like the questions in lines 3400 through 3490 you can easily
insert others.

Step 2 is used to enter the players as well as the sex
of each player. When doing this the computer makes sure
that an equal number of male and female players are partici-
pating (4500-4520).

Step 3 asks the questions and answers them. The answers are not displayed on the screen (Enter via GET, 5600-
5900).

Step 4 evaluates the answers and reports the results.
Every player answers the questions by entering numbers. 1
on the scale corresponds to the opinion "definitely no",
while 5 at the other end of the spectrum means "definitely
yes". From this point total the computer then calculates
the numerical difference in line 6200 and these differences

13

are then added: PM(i,k) = answer number of male i to question k; and PW(j,k) = answer number of female j to question k. An ideal match between male i and female j would be a sum S = 0. Complete incompatibility would be shown by S = 40 (using 10 questions). You can improve on this part of the program and its end results.

```
1000 : PRINT CHR$(147)
1010 : PRINT "          MATCH THE PARTNERS
1011 : PRINT "          _____
1012 :
1020 : PRINT
1030 : PRINT " I AM GOING TO ASK YOU A FEW QUESTIONS THAT
1040 : PRINT " YOU WILL BE ABLE TO ANSWER. YOUR ANSWERS WILL
1050 : PRINT " MAKE IT POSSIBLE TO MATCH YOU UP WITH SOMEONE
1060 : PRINT " IN THE ROOM AND TO RATE YOU ALL AS COUPLES.
1070 : PRINT
1090 :
1095 :
1100 REM   VARIABLES
1120 REM   M$(I)......MALE NAME CORRESPONDING TO I
1130 REM   F$(J)......FEMALE NAME CORRESPONDING TO J
1140 REM   Q$(K)......QUESTION CORRESPONDING TO K
1150 REM   PM(I,K)....ANSWER NUMBER OF MALE I TO QUESTION K
1160 REM   PW(J,K)....ANSWER NUMBER OF FEMALE J TO QUESTIONK
1170 REM   N..........MAXIMUM NUMBER OF COUPLES PLAYING
1900 :
1999 :
2000 REM*****MAIN PROGRAM*********************************
2001 :
2300 : GOSUB 3000 : REM INITIALISATION
2400 : GOSUB 4000 : REM ENTER NAME
2500 : GOSUB 5000 : REM QUESTION AND ANSWER
2600 : GOSUB 6000 : REM EVALUATION
2700 : GOSUB 7000 : REM EXIT
2900 : END
2970 :
2980 REM*******END OF MAIN PROGRAM************************
2995 :
2999 :
3000 REM ++++SUBROUTINE INITIALIZATION++++++++++++++++++++
3001 :
3100 : LET N = 10 : REM MAXIMUM NUMBER N COUPLES CAN PLAY.
3150 :
3200 : DIM M$(N), F$(N), Q$(10), PM(N,10), PF(N,10)
3300 :
3400 DATA DO YOU BELIEVE IN ASTROLOGY?
3410 DATA SHOULD WOMEN BE ALLOWED TO BECOME AIRLINE PILOTS?
3420 DATA ARE YOU GOOD AT REMEMBERING AND TELLING JOKES?
3430 DATA DO YOU SNORE?
```

14

```
3440 DATA IS NUCLEAR POWER A GOOD SOURCE OF ENERGY?
3450 DATA IS SMOKING UNHEALTHY?
3460 DATA DO YOU LIKE CLASSICAL MUSIC?
3470 DATA DO DOCTORS EARN TOO MUCH?
3480 DATA SHOULD MARRIED COUPLES TAKE SEPARATE VACATIONS?
3490 DATA DO YOU BELIEVE IN A SUPREME BEING?
3495 :
3500 : FOR K = 1 TO 10 : READ Q$(K) : NEXT K
3900 : RETURN
4000 REM ++++++SUBROUTINE ENTER NAME++++++++++++++++++++++++
4001 :
4010 : PRINT
4100 : PRINT " LET'S IDENTIFY THE PLAYERS.
4115 : PRINT
4200 : PRINT " PLEASE ENTER YOUR FIRST NAME AND SEX.
4210 : PRINT " AFTER THE LAST NAME, PLEASE TYPE 'END'.
4215 :
4300 : LET S = 1 : LET T = 1
4305 : LET M = 0 : LET F = 0
4310 : PRINT
4320 : INPUT "FIRST NAME"; N$
4325 : IF N$ = "END" THEN 4500
4330 : INPUT "SEX (M/F) "; S$
4340 : IF S$="M" THEN M$(S)=N$ : M=M+1 : S=S+1 : GOTO 4310
4350 : IF S$="F" THEN F$(T)=N$ : F=F+1 : T=T+1 : GOTO 4310
4360 : GOTO 4330
4400 :
4500 : IF M<F THEN PRINT "NEED ";F-M;" MEN" : GOTO 4310
4520 : IF F<M THEN PRINT "NEED ";M-F;" WOMEN" : GOTO 4310
4535 :
4540 : RETURN
4999 :
5000 REM ++++++SUBROUTINE QUESTION AND ANSWER ++++++++++++
5001 :
5005 : PRINT CHR$(147)
5100 : FOR I=1 TO M
5110 :    PRINT
5120 :    PRINT CHR$(18); M$(I)
5130 :    GOSUB 5500 : REM ANSWER KEY
5150 :    FOR K = 1 TO 10
5200 :      GOSUB 5600 : REM ANSWER
5250 :      LET PM(I,K) = VAL(A$)
5300 :    NEXT K
5320 : NEXT I
5340 :
5350 : PRINT CHR$(147)
5360 : FOR J = 1 TO W
5365 :    PRINT
5370 :    PRINT CHR$(18); W$(J)
5375 :    GOSUB 5500 : REM ANSWER KEY
5380 :    FOR K = 1 TO 10
5400 :      GOSUB 5600 : REM ANSWER
```
15

```
5450 :    LET PW(J,K)=VAL(A$)
5460 :   NEXT K
5470 : NEXT J
5480 :
5490 : RETURN : REM END SUBROUTINE "QUESTION AND ANSWER"
5499 :
5500 : REM ++++SUBROUTINE ANSWER KEY +++++++++++++++++++++++
5501 :
5510 : PRINT
5520 : PRINT "PLEASE ANSWER:
5525 : PRINT "(YOUR ANSWER WILL NOT APPEAR ON THE SCREEN.)"
5530 : PRINT
5540 : PRINT "DEFINITELY NO..............1"
5545 : PRINT "POSSIBLY NO................2"
5550 : PRINT "UNDECIDED .................3"
5555 : PRINT "POSSIBLY YES ..............4"
5560 : PRINT "DEFINITELY YES ............5"
5570 : PRINT
5590 : RETURN
5592 :
5595 : REM +++SUBROUTINE ANSWER++++++++++++++++++++++++++++++
5596 :
5600 : PRINT Q$(K) : REM COMPUTER'S QUESTION
5610 : GET  A$ : IF A$ = "" THEN 5610
5620 : IF A$="1" OR A$="2" OR A$="3" THEN RETURN
5630 : IF A$="4" OR A$="5" THEN RETURN
5670 : PRINT
5680 : PRINT "WRONG INPUT!"  : GOTO 5610
5900 :
6000 : ++++++SUBROUTINE SCORE++++++++++++++++++++++++++++++++
6001 :
6002 : PRINT CHR$(147)
6003 : PRINT "              RESULTS
6004 : PRINT "              _____
6006 : PRINT
6007 :
6100 : PRINT
6120 : FOR I = 1 TO M
6150 :   FOR J = 1 TO W
6160 :     LET S = 0
6180 :     FOR K = 1 TO 10
6200 :       LET D = ABS(PM(I,K)-PW(J,K))
6250 :       LET S = S+D
6300 :     NEXT K
6310 :    PRINT "COMPATIBILITY FACTOR FOR "; M$(I);
6320 :    PRINT " AND "; F$(J); "IS"; S
6340 :   NEXT J
6360 : NEXT I
6490 : PRINT   :   PRINT
6500 : PRINT "PRESS ANY KEY TO CONTINUE.
6550 : GET T$ : IF T$ = "" THEN 6550
6560 : RETURN
```

```
6900 :
7000 REM ++++++SUBROUTINE EXIT++++++++++++++++++++++++++++
7100 : PRINT CHR$(147)
7200 : PRINT : PRINT
7300 : PRINT "THANKS FOR ALL THOSE INTERESTING INSIGHTS!
7400 : PRINT
7500 : PRINT "GOOD-BYE.
7600 : PRINT : PRINT
7700 : FOR I = 1 TO 1000 : NEXT I
7800 : PRINT CHR$(147)
7900 : RETURN
```

BRAINTEASERS

1. Expand the program Addition Warm-ups so that the correct answer is displayed after three wrong answers and the player has to proceed to the next problem.

2. A further variation on the program Addition Warm-ups: at the end of a problem series the number of correct answers is displayed. Then the computer displays a comment to give the player a pat on the back, or else it displays a grade to evaluate the player's progress.

3. Build a time limit into the Addition Warm-ups. When a player oversteps this limit the computer interprets the problem as "unsolved."

4. The time limit can apply to the entire series of problems. Only those problems solved within the limit shall count.

5. Expand the program Addition Warm-ups by letting a player start by selecting various degrees of difficulty. Let the factors be either the size of the summands or the time allotted.

6. Design a program for Subtraction Warm-ups, Multiplication Warm-ups, and Division Warm-ups.

7. Mark the program Too High -- Too Low with a control to check the data entered. When a player enters a number lower than 1 or higher than the upper limit, the computer informs the player of this fact or rejects the data.

8. Design a program for a number guessing game called "Warmer -- Colder". It could either tell the player exactly how far off from the unknown number his guess was, or else

give clues about the discrepency by displaying a certain
number of symbols (e.g. ***).

9. Do not let the player in example 1.3 simply guess blind-
ly. After a specific number of attempts the computer should
interrupt the game. After how many guesses should it interrupt?

10. The winning numbers in the game Target 100 can also be
produced by a mathematical formula: $G=11*k+1$; $k=0,1...9$.
Try using this formula instead of the DATA line 810.

11. Write a program for Target 100 that makes the computer
into a game master instead of a opponent. The computer is
informed of the players' names and then calls on them when
their turn comes to play. The computer examines the entered
data (the number cannot be lower than 1 or higher than 10)
and also picks the winner.

12. Change the program of Target 100 so that several rounds
can be played. Let the computer display the number of
rounds won or lost at the end of each series (For example:
"The score is 5:3 for ... you").

13. Rewrite the program of Target 100 for other regions of
possible additions (i.e., but not from 1 - 10) and for other
"targets" ("Target 152"). Make it possible to choose the
target number and the region at the beginning.

14. The display of results in the game Match the Partners
might not be entirely satisfactory. Alter this so that your
screen displays a table or graph that looks something like
this:

	A	B	C	D
U:	20	12	18	24
X:	13	16	21	8
Y:	4	21	10	38
Z:	5	22	0	10

15. Try a further improvement on the Match the Partners program. Try having the computer independently name those people who, based on the numbers they have entered, seem mutually best suited for each other. These will be the people having lowest numbers in common. In the table above, male D is the best match with female X, male B goes with female U, etc.

16. You can also change the selection process by weighting various questions more heavily than others. Give your imagination free rein.

CHAPTER II

LEARNING GAMES

Ludendo discimus
(While playing, we learn.)

Leibnitz

The games in this chapter may strike some people as serious or useful while not being very playful. If you have to learn by these programs you may not find them that much fun, but on the other hand writing the programs offers all the more enjoyment. As an author of programs you might try to brighten your task of learning by inserting entertaining comments.

Example 2.1 THE ALPHABET GAME
Devise a program to teach children the alphabet.

Dialogue:

```
Enter the letters of the alphabet in correct order.
abcdefghik
I'm afraid that letter was wrong.  Please begin again
and enter the letters of the alphabet in correct order.
abcdefghijklmnopqrstuvwxyz
Well done!  Good-bye.
```

Here is the program for the Alphabet Game:

```
100 : PRINT CHR$(147)
110 : PRINT "          ALPHABET"
120 : PRINT "          --------"
130 :
140 REM THE OBJECT IS TO TEACH CHILDREN OVER AGE 3 THE
145 REM ALPHABET.
150 :
160 : PRINT
170 : PRINT "ENTER THE LETTERS OF THE ALPHABET IN ORDER
180 : PRINT
190 :
```

```
200 : FOR N = 65 TO 90 : REM ASCII NUMBERS OF THE LETTERS.
210 :
220 :    GET B$ : IF B$ = "" THEN 220 : REM LETTER ENTRY.
230 :    PRINT B$;"  ";
240 :
250 :    IF B$ = CHR$(N) THEN 320 : REM CORRECT, NEXT LETTER.
260 :    PRINT : PRINT
270 :    PRINT "SORRY, THE LAST LETTER WAS WRONG.";
280 :    PRINT "START AGAIN FROM THE BEGINNING!
290 :    PRINT : PRINT
300 :    GOTO 170 : REM JUMP TO BEGINNING
310 :
320 : NEXT N
330 :
340 : PRINT : PRINT
350 : PRINT "WELL DONE! GOOD-BYE.
360 :
370 : END
```

This program is quite simple. The counting variable N in line 200 runs from 65 through 90. These numbers correspond to the letters A - Z in the ASCII code, which stands for American Standard Code for Information Interchange (consult the user's manual for your computer for more information on this). When an incorrect letter is entered, the program begins again and doesn't let the player exit until the entire alphabet has been entered correctly.

Now here is a game to train your memory and concentration:

Example: 2.2 NUMBER SENSOR
The computer displays some randomly chosen numbers which the player must repeat. After each successful repetition the string increases in length by one number.

Dialogue:

```
Current number string: 4
Repeat the string: 4
Current number string: 4 7
Repeat the string: 4 7
Current number string: 4 7 1
Repeat the string: 4 7 2
Sorry, that's wrong!
The correct string is: 4 7 1
```

The process looks like this:

NUMBER SENSOR

```
REPEAT
   Simultaneously generate and store random numbers
   Display the previous number sequence
   Enter the number sequence of the player
UNTIL  wrong number entered
END REPEAT
```

In the first part of the program that follows, a whole number is generated at random (line 240) and entered into a table. In the second part, all numbers previously entered are flashed on the screen (the display time is determined by the waiting loop in line 330). In the third part the user must repeat the number sequence. If a wrong number is entered, the program moves to part four (Break Off); otherwise it jumps back to the first part.

```
0100 : PRINT CHR$(147)
0110 : PRINT "     NUMBER SENSOR
0120 : PRINT "     -------------
0130 :
0140 REM GAME TO TRAIN CONCENTRATION AND MEMORY
0150 :
0160 : DIM A(100) : REM TABLE TO STORE DIGITS
0170 :
0180 REM === GENERATING THE NUMBERS =====================
0190 :
0200 : PRINT
```

```
0210 : LET N = 1                            : REM NUMBER COUNTER
0220 : PRINT : PRINT
0230 : PRINT N;". ATTEMPT: "
0240 : LET Z = INT(9*RND(TI)+1)     : REM RANDOM NUMBER
0250 : LET A(N) = Z                 : REM STORING THE NUMBERS
0260 :
0270 REM === PRESENTATION OF THE NUMBER ORDER ==============
0280 :
0290 : PRINT
0300 : PRINT "ORDER OF NUMBERS TO THIS POINT: ";
0310 : FOR I = 1 TO N : PRINT CHR$(18); A(I);: NEXT I
0320 :
0330 : FOR J = 1 TO 500 : NEXT J         : REM WAITING LOOP
0340 : PRINT CHR$(147)
0350 :
0360 REM === REPEAT FROM MEMORY ===========================
0370 :
0380 : PRINT
0390 : PRINT "ENTER THE NUMBERS IN ORDER:"
0400 : FOR I = 1 TO N
0410 :    GET R$ ; IF R$ = "" THEN 410        : REM ENTER
0420 :    IF R$ < "1" OR R$ > "9" THEN 410    : REM CHECK
0430 :    LET R = VAL(R$)                 : REM CHANGE TO NUMBER
0440 :    PRINT R;
0450 :    IF R <> A(I) THEN 520           : REM MISREMEMBERED
0460 : NEXT I
0470 :
0480 : PRINT : PRINT "CORRECT!"
0490 :
0500 : LET N = N+1 : GOTO 220                 : REM JUMP BACK
0510 :
0520 REM === BREAK OFF ====================================
0530 :
0540 : PRINT : PRINT
0550 : PRINT "SORRY, INCORRECT.  THE CORRECT ORDER IS:
0560 : PRINT
0570 : FOR I = 1 TO N : PRINT A(I);: NEXT I
0580 :
0590 : PRINT : PRINT
0600 :
0610 : END
```

Example 2.3 TOTAL RECALL

Related words (such as names or concepts) appear in randomly
chosen positions on the video screen. The player must remember these and re-enter them.

A possible dialogue:

```
crow
                            hawk

    sparrow
                thrush
```

After a wait of N seconds:

```
Now enter the same words.
    sparrow    Correct!
    bluejay    Sorry, that's incorrect.
    crow       Correct!
Here are the words you were supposed to memorize:
    crow      hawk      sparrow      thrush
These are the ones you memorized:
    sparrow    crow
```

There are five parts to this program: initialization, rules
of play, display of words to be memorized, player entry,
and evaluation. In the first part we prepare a table B$ of
up to 100 words which is filled in the DATA lines 880 and
following. Then the computer prints the rules if the player
so desires. In the third part it first finds a random cate-
gory Z (line 465) and proceeds to determine the related word
or concept (line 480). Then, with X and Y, it selects a
random spot on the screen and marks this with the cursor.
(The cursor movements in lines 520-545 are covered in the
following chapter.) The chosen word is then printed in
this spot. The process is repeated D times (in the program
D = 8, so eight words are displayed). In the fourth part of
the program the player enters the words remembered. Finally, the computer displays all the words the player originally saw as well as those that the player remembered.

```
00100 : PRINT CHR$(147)
00110 : PRINT "          TOTAL RECALL"
00120 : PRINT "          ------------"
00130 :
00140 REM GAME TO TRAIN MEMORY AND QUICKNESS OF MIND
00150 :
00160 REM === INITIALIZATION ============================
```

```
00170 :
00180 : DIM W$(100)                : REM TABLE OF WORDS
00190 : LET D = 8                  : REM NUMBER OF WORDS LISTED
00200 : LET T = 200                : REM TIME CONSTANT
00210 :
00215 : DIM N$(D), M$(D)
00220 : FOR I = 1 TO 100
00230 :   READ W$
00240 :   IF W$ = "@" THEN 270 : REM ALL WORDS ENTERED
00250 :   LET W$(I) = W$
00260 : NEXT I
00270 : LET N = I-1               : REM NUMBER OF ALL WORDS
00280 :
00290 REM === RULES OF THE GAME ============================
00300 :
00310 : PRINT
00320 : PRINT "DO YOU WANT TO SEE THE RULES? (Y/N)
00330 : GET A$ : IF A$ = "" THEN 330
00340 : IF A$ <> "Y" THEN 440 : REM SKIP GAME RULES
00350 :
00360 : PRINT : PRINT
00370 : PRINT "YOU WILL SEE A SERIES OF RELATED WORDS
00380 : PRINT "YOUR OBJECT IS THEN TO ENTER AS MANY OF THESE
00390 : PRINT "AS YOU CAN REMEMBER.
00400 :
00410 : PRINT : PRINT
00420 : PRINT "PRESS ANY KEY!
00430 : GET K$ : IF K$ = "" THEN 430
00435 :
00440 REM === DISPLAY WORDS ================================
00450 :
00460 : PRINT CHR$(147)
00465 : LET Z = INT((N/D)*RND(TI))    : REM RANDOM CATEGORY
00470 : FOR J = 1 TO D
00480 :   LET K = J+8*Z                       : REM WORD
00490 :   LET M$(J) = W$(K)            : REM REMEMBER WORD
00500 :   LET X = INT(39*RND(TI)+1)          : REM COLUMN
00510 :   LET Y = INT(24*RND(TI)+1)          : REM ROW
00520 :   PRINT CHR$(19);             : REM CURSOR BACK HOME
00540 :   FOR R = 1 TO X : PRINT CHR$(29); : NEXT R
00542 :     REM CURSOR MOVES RIGHT
00545 :   FOR R = 1 TO Y : PRINT CHR$(17); : NEXT R
00547 :     REM CURSOR MOVES DOWN
00550 :   PRINT M$(J)                      : REM PRINT WORD
00560 : NEXT J
00570 :
00580 : FOR R = 1 TO T : NEXT R          : REM WAITING LOOP
00585 : PRINT CHR$(147)
00590 :
00600REM === PLAYER ENTRY =================================
00610 :
00620 : PRINT
```

```
00630 : PRINT "REPRODUCE THE WORDS."
00640 : PRINT "WHEN YOU HAVE ENTERED THEM ALL, PRESS '@'!
00650 : PRINT
00660 :
00670 : FOR J = 1 TO D
00680 :   INPUT B$
00690 :   IF B$ = "@" THEN 770      : REM ALL WORDS ENTERED
00700 :   FOR C = 1 TO 15 : PRINT CHR$(29); : NEXT C
00703 :     REM CURSOR RIGHT
00705 :   PRINT CHR$(145);          : REM CURSOR UP
00710 :   FOR L = 1 TO D
00720 :     IF B$ <> M$(L) THEN 730
00725 :     PRINT "CORRECT!" : N$(L) = B$ : GOTO 750
00730 :   NEXT L
00740 :   PRINT "SORRY, THAT'S WRONG.
00750 : NEXT J
00760 :
00770REM === EVALUATION =====================================
00780 :
00790 : PRINT : PRINT : PRINT
00800 : PRINT "THESE WERE THE WORDS:
00810 : PRINT
00820 : FOR J = 1 TO D : PRINT M$(J);" "; : NEXT J
00830 : PRINT : PRINT : PRINT
00840 : PRINT "AND THESE ARE THE ONES YOU REMEMBERED:
00850 : PRINT
00860 : FOR J = 1 TO D : PRINT N$(J);" "; : NEXT J
00870 :
00880 DATA DOG, CAT, MOUSE, BIRD, WORM, RAT, HORSE, COW
00890 DATA BACH, VERDI, BRAHMS, MOZART, MENDELSSOHN
00895 DATA BERLIOZ, DVORAK, HANDEL, BEECH, HEMLOCK, PINE
00900 DATA MAPLE, ELM, OAK, HICKORY, SPRUCE, TORONTO
00910 DATA HARTFORD, MIAMI, TUCSON, SACRAMENTO, BANGOR
00915 DATA BOSTON, TULSA, @
00920 DATA YOU CAN FIT MANY MORE HERE. REMEMBER, THE LAST
00930 DATA CHARACTER IS ALWAYS @
```

This game is another one to train your powers of memory and concentration:

Example 2.4 CARD SHARK'S MEMORY

A deck is shuffled, the tens are removed and
the cards laid on the table face down. The
player whose turn it is turns over two cards.
If they have the samenumerical (or face)
value, then he can remove them from the table
and turn over two more cards. Otherwise, he
replaces the two he turned over and the next
player takes a turn. The player who has removed the most
cards from the table by the end of the game is the winner.
The challenge is to remember which cards have been turned
over during the game and where they lie on the table.

This is what a round looks like when simulated on the computer:

```
It is Tom's turn.
What cards are you turning over?
     1  2  3  4  5  6  7  8  9 10 11 12
1    X  X  X  X  X  X  X  X  X  X  X  X
2    X  X  X  X  X  X  X  X  X  X  X  X
3    X  X  X  X  X  X  X  X  X  X  X  X
4    X  X  X  X  X  X  X  X  X  X  X  X
FIRST CARD? 1,2
SECOND CARD? 4,7
     1  2  3  4  5  6  7  8  9 10 11 12
1    X  6♥ X  X  X  X  X  X  X  X  X  X
2    X  X  X  X  X  X  X  X  X  X  X  X
3    X  X  X  X  X  X  X  X  X  X  X  X
4    X  X  X  X  X  X  6♦ X  X  X  X  X
Very Good!
```

```
00100 : PRINT CHR$(147)
00110 : PRINT "           CARD SHARK'S MEMORY
00120 : PRINT "           ------------------
00130 :
00140 REM GAME TO TRAIN MEMORY AND CONCENTRATION
00150 :
00160 :
00170 REM *** MAIN PROGRAM *******************************
```

28

```
00180 :
00190 : GOSUB 460 : REM RULES OF PLAY
00200 : GOSUB 570 : REM INITIALIZATION
00210 : LET S = 0 : REM PLAYER 'SWITCH'
00220 :
00230 : FOR C = 1 TO 24
00240 :
00250 :    PRINT CHR$(147)
00260 :    PRINT : PRINT "PLAYER" CHR$(18) N$(S) CHR$(146);
00265 :    PRINT "TAKES A TURN!";
00270 :    PRINT "WHAT CARDS DID YOU TURN OVER?
00280 :    LET X = 0 : LET Y = 0 : LET I = 0 : LET V = 0
00290 :
00300 :    GOSUB 1100 : REM DISPLAY FIELD OF PLAY
00310 :    GOSUB  960 : REM ENTER CARDS
00320 :    GOSUB 1100 : REM DISPLAY FIELD OF PLAY
00330 :    GOSUB 1300 : REM TEST FOR IDENTICAL CARDS
00340 :
00350 :    FOR W = 1 TO 1500 : NEXT W : REM WAITING LOOP
00360 :
00370 :    IF T$ <> "MATCH" THEN LET S = 1-S : GOTO 250
00380 :
00390 : NEXT C
00400 :
00410 : END
00420 :
00430 REM *** END OF MAIN PROGRAM *************************
00440 :
00450 :
00460 REM +++ SUBROUTINE RULES OF PLAY ++++++++++++++++++++
00470 :
00480 : PRINT
00490 : PRINT "WOULD YOU LIKE TO SEE THE GAME RULES? (Y/N)
00500 : GET A$ : IF A$ = "" THEN 500
00510 : IF A$ <> "Y" THEN RETURN
00520 :
00530 : REM --- INSERT THE RULES OF PLAY HERE --------------
00540 :
00550 : RETURN
00560 :
00570 REM +++ SUBROUTINE INITIALIZATION +++++++++++++++++++
00580 :
00590 : PRINT : PRINT "WHO IS PLAYING?
00600 : PRINT
00610 : INPUT "FIRST PLAYER"; N$(0)
00620 : INPUT "SECOND PLAYER"; N$(1)
00630 :
00640 : LET T(0) = 0 : LET T(1) = 0 : REM SCORE
00650 :
00660 : PRINT : PRINT "PLEASE BE PATIENT";
00670 : PRINT "I'M SHUFFLING THE CARDS.
00680 :
```

```
00690 : REM --- PREPARE THE CARDS -------------------------
00700 :
00710 : DIM F$(4,12)
00720 :
00730 : FOR I = 1 TO 4
00740 :   FOR J = 1 TO 12 : READ F$(I,J) : NEXT J
00750 : NEXT I
00760 :
00770 : REM --- SHUFFLE CARDS ----------------------------
00780 :
00790 : FOR I = 1 TO 4
00800 :   FOR J = 1 TO 12
00810 :     LET X = INT(4*RND(TI)+1)
00820 :     LET Y = INT(12*RND(TI)+1)
00830 :     LET H$ = F$(I,J)
00840 :     LET F$(I,J) = F$(X,Y)
00850 :     LET F$(X,Y) = H$
00860 :   NEXT J
00870 : NEXT I
00880 :
00890 DATA  "2♦","3♦","4♦","5♦","6♦","7♦","8♦"
00895 DATA  "9♦","J♦","Q♦","K♦","A♦"
00900 DATA  "2♥","3♥","4♥","5♥","6♥","7♥","8♥"
00905 DATA  "9♥","J♥","Q♥","K♥","A♥"
00910 DATA  "2♠","3♠","4♠","5♠","6♠","7♠","8♠"
00915 DATA  "9♠","J♠","Q♠","K♠","A♠"
00920 DATA  "2♣","3♣","4♣","5♣","6♣","7♣","8♣"
00925 DATA  "9♣","J♣","Q♣","K♣","A♣"
00930 :
00940 : RETURN
00950 :
00960 REM +++ SUBROUTINE INPUT +++++++++++++++++++++++++++++
00970 :
00980 : PRINT
00990 : INPUT "FIRST CARD "; X,Y
01000 : IF X < 0 OR X > 4 OR Y < 0 OR Y > 12 THEN 990
01010 : IF F$(X,Y) <> " " THEN 1030
01015 : PRINT "THAT CARD IS GONE." : GOTO 990
01020 :
01030 : INPUT "SECOND CARD "; U,V
01040 : IF U < 0 OR U > 4 OR V < 0 OR V > 12 THEN 1030
01050 : IF U = X AND V = Y THEN 1030
01060 : IF F$(U,V) <> " " THEN 1080
01065 : PRINT "THAT CARD IS GONE." : GOTO 1030
01070 :
01080 : RETURN
01090 :
01100 REM +++ SUBROUTINE DISPLAY FIELD OF PLAY ++++++++++++
01110 :
01120 : PRINT : PRINT
01130 : FOR J = 1 TO 12 : PRINT TAB(3*J+3) J; : NEXT J
01140 : PRINT : PRINT
```

```
01150 :
01160 : FOR I = 1 TO 4
01170 :    PRINT I;
01180 :    FOR J = 1 TO 12
01190 :       IF I = X AND J = Y    THEN 1230
01200 :       IF I = U AND J = V    THEN 1230
01210 :       IF F$(I,J) = " "       THEN 1230
01220 :       PRINT TAB(3*J+3) "▓"; : GOTO 1240
01230 :       PRINT TAB(3*J+3) F$(I,J);
01240 :    NEXT J
01250 :    PRINT
01260 : NEXT I
01270 :
01280 : RETURN
01290 :
01300 REM +++ SUBROUTINE TEST FOR MATCHING CARD ++++++++++++
01310 :
01320 : IF LEFT$(F$(X,Y),1) = LEFT$(F$(U,V),1) THEN 1370
01330 :
01340 : PRINT
01350 : PRINT "SORRY, THAT'S WRONG." : LET T$ = "N"
01355 : GOTO 1420
01360 :
01370 : PRINT
01380 : PRINT "IT MATCHES!" : LET T$ = "MATCH"
01390 : LET F$(X,Y) = " " : LET F$(U,V) = " "
01395 :    REM REMOVE CARDS
01400 : LET T(S) = T(S) + 1 : REM INCREASE SCORE
01410 :
01420 : PRINT : PRINT "SCORE SO FAR:"
01430 : PRINT CHR$(18) N$(0); ":"; T(0)
01440 : PRINT CHR$(18) N$(1); ":"; T(1)
01450 :
01460 : RETURN
```

Explanation:

This program is written for two players. Changing between the two players is made possible by the variable S which can have the values 0 or 1. At the beginning S = 0 (line 210), thus it is the turn of the player named N$(0) (line 260). If this player turns over two cards with different values, the computer executes S = 1-S, meaning that now S = 1 (line 370) and N$(1) takes a turn (line 260). If a player scores, the program proceeds to the next value of C (line 390). Because there are 24 pairs of cards possible, the variable C counts from 1 to 24 (line 230). The cards that match are removed from play (line 1390) so they are not displayed again (line 1210).

Here is a program that is a serious teaching aid. It will help you improve your math with whole numbers. It is an expanded version of example 1.2 in the preceding chapter.

Example 2.5 MATH PRACTICE
The computer asks the player a variety of problems in addition, subtraction, multiplication, and division using whole numbers between 1 and 100. The player has to supply the answer.

```
01000 : PRINT CHR$(147)
01100 : PRINT "              MATH PRACTICE
01200 : PRINT "              -------------
01300 :
01400 REM PRACTICE IN THE FOUR BASIC MATHEMATICAL OPERATIONS
01500 :
01600 REM *** MAIN PROGRAM ********************************
01700 :
01800 : PRINT
01900 : INPUT "HOW MANY PROBLEMS DO YOU WANT TO TRY";N
02000 : PRINT
02100 : PRINT "DEGREE OF DIFFICULTY E (EASY), M (MED), H";
02110 : PRINT " (HARD) ?"
02200 : GET A$ :   IF A$ = "" THEN 2200
02300 :
02400 : IF A$ = "E" THEN LET W = 10 : GOTO 2900
02500 : IF A$ = "M" THEN LET W =  6 : GOTO 2900
02600 : IF A$ = "H" THEN LET W =  3 : GOTO 2900
02610 :    REM W SECONDS PER PROBLEM
02700 : GOTO 2200
02800 :
02900 : LET L = N*W*60              : REM TIME LIMIT
03000 :
03100 : LET T = TI                  : REM SETTING THE CLOCK
03200 :
03300 : LET R = 0                   : REM NUMBER CORRECT
03400 :
03500 : DEF FNR(X) = INT(8*RND(X)+2)  : REM RANDOM FUNCTION
03600 : DEF FNS(X) = INT(89*RND(X)+11): REM RANDOM FUNCTION
03700 :
03800 : REM --- LOOP START ---------------------------------
03900 :
04000 : FOR I = 1 TO N
04100 :
04200 : REM --- CHOICE OF OPERATION ---
04300 :
04400 :    LET K = INT(4*RND(1)+1)
04500 :    ON K GOSUB 7800, 8700, 9600, 10500
```

```
04600 :
04700 : REM --- QUESTION AND ANSWER ---
04800 :
04900 :    PRINT : PRINT "HOW MUCH IS "; Z1; OP$; Z2; " ";
05000 :    INPUT A
05100 :
05200 :    IF TI-T < L THEN 5250
05210 :    PRINT : PRINT "TIME IS UP!" : GOTO 5900
05250 :    IF A=E THEN PRINT "RIGHT!" : LET R=R+1:   GOTO 5500
05300 :    PRINT "SORRY, THAT'S WRONG."
05400 :
05500 : NEXT I
05600 :
05700 : REM --- LOOP END ----------------------------------
05800 :
05900 : REM --- SCORE ------------------------------------
06000 :
06100 : LET Q = INT(1000*R/N+0.5)/1000
06200 : PRINT : PRINT
06300 : PRINT "OUT OF ";N;" PROBLEMS YOU GOT "R" CORRECT.";
06400 : PRINT "THAT IS "; 100*Q; " PERCENT."
06500 :
06600 : REM --- REPEAT OR EXIT ---------------------------
06700 :
06800 : PRINT : PRINT : PRINT
06900 : PRINT "ANOTHER SERIES? (Y/N)
07000 : GET A$ : IF A$ = "" THEN 7000
07100 : IF A$ = "Y" THEN RUN
07200 : PRINT: PRINT: PRINT "GOOD-BYE."
07300 :
07400 : END
07500 :
07600 REM *** END OF MAIN PROGRAM **************************
07700 :
07800 REM +++ SUBROUTINE ADDITION +++++++++++++++++++++++++++
07900 :
08000 : LET OP$ = "+"
08100 : LET Z1 = FNS(1)
08200 : LET Z2 = FNS(1)   : REM CHOICE OF SUMMANDS
08300 : LET E = Z1 + Z2        : REM RESULT
08400 :
08500 : RETURN
08600 :
08700 REM +++ SUBROUTINE SUBTRACTION +++++++++++++++++++++++++
08800 :
08900 : LET OP$ = "-"
09000 : LET Z1 = FNS(1)
09100 : LET Z2 = FNS(1)  : REM CHOICE OF OPERANDS
09200 : LET E = Z1 - Z2        : REM RESULT
09300 :
09400 : RETURN
09500 :
```

```
09600 REM +++ SUBROUTINE MULTIPLICATION +++++++++++++++++++++
09700 :
09800 : LET OP$ = "*"
09900 : LET Z1 = FNR(1)        : REM FIRST FACTOR - ONE DIGIT
10000 : LET Z2 = FNS(1)        : REM SECOND FACTOR - TWO DIGITS
10100 : LET E = Z1 * Z2        : REM RESULT
10200 :
10300 : RETURN
10400 :
10500 REM +++ SUBROUTINE DIVISION +++++++++++++++++++++++++++++
10600 :
10700 : LET OP$ = "/"
10800 : LET Z2 = FNR(1)        : REM ONE DIGIT DIVISOR
10900 : LET E = FNS(2)         : REM TWO DIGIT QUOTIENT
11000 : LET Z1 = E/Z2          : REM DIVIDEND
11100 :
11200 : RETURN
11300 :
11400 : END
```

Explanation:

Lines 4400 and 4500 of the program choose one of the four basic areas of mathematical calculation. At first the computer generates one of the numbers 1,2,3,4 at random. In conjunction with this number it then jumps into the subroutine for addition, subtraction, multiplication or division. The level of difficulty is chosen in lines 2400-2600: $L=N*W*60$ is the time allotted for solving N problems. The clock is set in line 3100: TI (= time) is a variable governing time elapsed since power-on in 60ths of a second. Line 5200 tests whether the player goes over the time limit L.

Example 2.6

STATE YOUR TERMS

An algebraic term is displayed and must be restated. The computer tells you whether your restatement is an equivalent of the term displayed.

$(a+b)*(c+d)$

34

Dialogue:

```
Enter the term.
  (a+b) ↑ 2
Equivalent term?
 a↑2+2*a*b+b↑2
Correct!
Equivalent term?
@
Good-bye!
```

Here is the program:

```
00100 : PRINT CHR$(147)
00110 : PRINT "      STATE YOUR TERMS
00120 : PRINT "      ----------------
00130 :
00140 REM INSTRUCTIONAL PROGRAM TO TEACH ALGEBRA
00150 :
00160 : PRINT
00170 : PRINT "YOU CAN ENTER AND MODIFY A TERM MADE UP OF
00180 : PRINT "NUMBERS, LETTERS, AND OPERATIONAL SYMBOLS
00185 : PRINT " +,-,*,/,↑ ,( ).
00190 : PRINT "THEN I'LL TELL YOU IF THE MODIFIED
00200 : PRINT "TERM IS EQUIVALENT TO THE ONE GIVEN.
00210 : PRINT "WHEN YOU'VE HAD ENOUGH, PRESS @.
00220 :
00230 REM *** MAIN PROGRAM ********************************
00240 :
00250 : GOSUB 450  : REM ASSIGNING RANDOM NUMBERS
00260 :
00270 : PRINT
00280 : INPUT "WHAT TERM DO YOU WANT TO MODIFY"; A$
00290 : PRINT
00300 : PRINT "TERM TO BE MODIFIED: "; A$
00310 : PRINT
00320 : INPUT " EQUIVALENT TERM: "; B$
00330 : IF B$ = "@" THEN END
00340 :
00350 : GOSUB 700 : REM CALCULATING THE TERM VALUE
00360 :
00370 : PRINT
00380 : PRINT "THE MODIFICATION IS ";
00390 :
00400 : IF ABS(DF) > 0.001 THEN PRINT "WRONG" : GOTO 290
00410 : PRINT "CORRECT!! " :  LET A$ = B$ : GOTO 290
00420 :
00430 REM *** END OF MAIN PROGRAM *************************
```

```
00440 :
00450 REM +++ SUBROUTINE 'ASSIGNING RANDOM NUMBERS' ++++++++
00460 :
00470 : PRINT : PRINT
00480 : PRINT "NOW YOU'LL SEE THE ASSIGNING OF RANDOM";
00490 : PRINT " NUMBERS TO THE VARIABLES.
00500 : PRINT "PRESS ANY KEY.
00510 : GET A$ : IF A$ = "" THEN 510
00520 :
00530 : FOR I1 = ASC("A") TO ASC("Z")
00540 :    PRINT CHR$(147);CHR$(29);CHR$(29);CHR$(I1);" = ";
00545 :    PRINT RND(0);" : GOTO 570
00550 :    GOSUB 620 : REM FILLING THE KEYBOARD BUFFER
00560 :    END
00570 : NEXT I1
00580 : POKE 158, 0 : REM NO CHARACTER IN KEYBOARD BUFFER
00590 :
00600 : RETURN
00610 :
00620 REM +++ SUBROUTINE FILLING THE KEYBOARD BUFFER +++++++
00630 :
00640 : POKE 623, ASC(CHR$(19))  : REM CURSOR HOME
00650 : POKE 624, 13            : REM RETURN
00660 : POKE 158, 2   : REM TWO CHARACTERS IN KEYBOARD BUFFER
00670 :
00680 : RETURN
00690 :
00700 REM +++ SUBROUTINE CALCULATING THE TERM VALUE ++++++++
00710 :
00720 : PRINT CHR$(147) "DF = (";A$;") - (";B$;") : GOTO 750
00730 : GOSUB 620 : REM ASSIGNING THE KEYBOARD BUFFER
00740 : END
00750 : FOR I = 32768 TO 32768+239 : POKE I,32 : NEXT I
00760 : PRINT
00770 :
00780 : RETURN
```

Explanation:

The variables A through Z are given random numbers between 0 and 1. After each attempted restatement of the term, the computer evaluates the terms using the random numbers and comes up with an original answer and a new answer. The computer compares the two answers. If they are identical then the terms are most probably equivalent; otherwise definitely not. The computer supplies these responses and challenges the player to try new restatements of the term.

The following points are a little tricky and will be explained fully in chapter 3. We want to write a program that can evaluate the term immediately without interrupting the program. Normally, terms are part of the program. In changing them you would change the program and thereby have to start over again at the beginning. Let's keep the program running without changes. To do this we use the fact that the END command lets the computer enter the direct mode and then gets the number of characters from the keyboard buffer (memory cells 623-632) given in the buffer counter (cell 158), and then executes them. If we display a command on the screen before the end of the program and put the cursor at the beginning of this line, then the computer will continue with this line after executing an END. In order to do this we must enter in the keyboard buffer the spot where we want the cursor to move to as well as the command RETURN.

This is precisely what makes our program. For the letters A through Z we write the following in line 540 and display it on the screen:

(*)   LET letter = random number   : GOTO 570

The subroutine "Filling the Keyboard Buffer" places the cursor in the upper left corner after the END command where it finds the command (*). Then it receives from the keyboard buffer the command RETURN (ASCII number 13). This causes the completion of the command (*); that is, the letter is assigned a random number and the computer proceeds with line 570 (NEXT I1).

The difference DF between the terms A$ (the original term) and B$ (the restated term) is then calculated in line 720, using the same routine.

As a final example in this chapter we shall develop a program that will be a transition to games with graphics.

37

Example 2.7 DRIVER'S ED

The computer simulates a road that winds
in random curves and turns across the screen.
The player has to keep the car in the middle
without hitting the shoulders.

Let's play the game as simply as possible
without recourse to the techniques of computer
graphics that we are going to learn in the
next chapter. The road comes into view at
the bottom of the screen which gives the
impression of driving backwards. To
drive the car (marked by +) we use a
variable T$. If the player does not
touch the keys 1 or 2, the computer waits
a moment (line 370) and drives the car
perpendicularly downwards (maintaining
the direction). Random numbers change
the road (line 490) and these determine
which of the characters /, I, or \ are
used to mark the contours. When the
car hits the shoulder, we get CRASH!
(line 680) and the game begins again.

```
00100 : PRINT CHR$(147)
00110 : PRINT "           DRIVER'S ED"
00120 : PRINT "           -----------
00130 :
00140 REM SIMPLE GAME OF SKILL
00150 :
00160 : PRINT "YOU'LL STEER YOUR CAR  '+'  WITH THE KEYS
00170 : PRINT "1 (TO THE LEFT) AND 2 (TO THE RIGHT)
00180 :
00190 : PRINT : PRINT "PRESS ANY KEY!
00200 : GET A$ : IF A$ = "" THEN 200
00210 :
00220 REM --- INITIALIZATION ------------------------------
00230 :
00240 : LET B = 10      : REM WIDTH OF ROAD
00250 : LET W = 100     : REM TIME CONSTANT FOR WAIT LOOP
00260 :
00270 : LET LR = 20     : REM LEFT ROAD EDGE
00280 : LET RR = LR + B : REM RIGHT ROAD EDGE
```

```
00290 :
00300 : LET P = 24     : REM STARTING POSITION OF CAR
00310 : LET M = 0      : REM POINT TOTAL
00320 :
00330 REM --- STEERING THE CAR -----------------------------
00340 :
00350 : GET T$
00360 : IF T$ <> "" THEN 380 : REM STEERING KEY WAS PRESSED
00370 : FOR I = 1 TO W : NEXT I : GOTO 430 : REM HESITATION
00380 : IF NOT(T$ = "1" OR T$ = "2") THEN 350
00390 :
00400 : IF T$ = "1" THEN LET P = P-1 : GOTO 430
00405 :    REM CAR GOES LEFT
00410 : LET P = P+1 : REM CAR GOES RIGHT
00420 :
00430 : IF P = LR OR P = RR THEN 660  : REM INTO THE DITCH
00440 :
00450 : LET M = M+1                : REM RAISE POINT COUNTER
00460 :
00470 REM --- RANDOM MOVEMENT OF ROAD ----------------------
00480 :
00490 : LET X = INT(3*RND(TI))-1
00500 :
00510 : IF X = -1 THEN LET D$ = "/" : GOSUB 550 : GOTO 350
00520 : IF X = +1 THEN LET D$ = "\" : GOSUB 550 : GOTO 350
00530 :                LET D$ = "I" : GOSUB 550 : GOTO 350
00540 :
00550 REM +++ SUBROUTINE 'CHANGING COORDINATES OF ROAD'+++++
00560 :
00570 : LET LR = LR + X                 : REM LEFT EDGE
00580 : IF LR < 1 THEN LET LR = 1       : REM CHECK LEFT
00590 : LET RR = LR + B                 : REM RIGHT EDGE
00600 : IF RR > 39 THEN LET LR = 29 : GOTO 590
00605 :                                 : REM CHECK RIGHT
00610 :
00620 : PRINT TAB(LR) D$; TAB(P) "+"; TAB(RR) D$
00625 :    REM SHOW ROAD
00630 :
00640 : RETURN
00650 :
00660 REM --- SCORE ----------------------------------------
00670 :
00680 : PRINT TAB(P)  "            CRASH!!"
00690 : PRINT : PRINT
00700 : PRINT "YOU GOT A TOTAL OF "M" POINTS.
00710 :
00720 : FOR I = 1 TO 1500 : NEXT I
00730 : PRINT CHR$(147) : GOTO 220     : REM GAME STARTS OVER
```

BRAINTEASERS

1. Expand the program Number Sensor in such a way as to place a limit on the amount of time a player has. Make this a variable and let the player select it as a degree of difficulty in advance.

2. In the game Number Sensor add a score and comment on a player's guessing performance (length of the number series when game ends).

3. Devise a program that will test whether the user can type in the name of the day of the week as well as the months.

4. Write a program that tests whether a user can supply the correct integer that follows the one displayed by the computer. For example, "What number follows 9909?"

5. In the program Total Recall a word or concept may partially cover another word. Correct this problem.

6. Let players of Total Recall select in advance the number of words displayed as well as the time to memorize them.

7. In the program Total Recall the words appear on the screen one after another. Customize the program so that the words and their place on the screen can be selected in advance. Write the program so that the words appear together instantaneously on the screen.

8. Complete the program Card Shark's Memory by adding the rules of play and a display of the game's final outcome showing the cards after the last move.

9. In Card Shark's Memory the playing area can move around on the screen. Create an immovable playing area.

10. Expand Card Shark's Memory so that several players can play.

11. Change the random law governing the road contours in Driver's Ed so it is possible to have longer straight stretches and curves.

12. The computer shall randomly generate a number N of letters and present these to the player to memorize. You should be able to select the number of letters in advance. You must memorize them within a certain time limit. The computer examines the letters in the response to see if they are correct and then gives a score.

13. Write a program to help a user learn vocabulary words.

14. SPEED READING. A sentence appears briefly on the screen. The player must remember and retype the sentence correctly.

15. FRACTION PRACTICE. Write a program analagous to Addition Warm-ups that presents problems with fractions and checks players' answers.

16. NUMBER TRANSFORMATION. Devise a program that tests the user's ability to transform numbers from the decimal system to the binary system and vice versa. Dialogue example:

```
Binary number: 11000
Decimal?   24
Correct!
Decimal number: 7
Binary?   1111
Sorry. Incorrect.
```

17. ROUGH ESTIMATE. Devise a program that gives practice in estimating answers for the four basic mathematical operations as well as for finding square roots.

Example:

```
1.85 * 2132 = ?   4000
Good! Your guess is off by 1.4%
Square root of 230?   15
Margin of Error: 1.1%
```

The program must ensure that a maximum of two appropriate numbers are entered. For example, the computer would reject the answer 3995 because you couldn't have arrived at it by rough calculation.

18. QUADRATIC EQUATIONS. Write a program that generates a solution to a quadratic equation. For example:

```
x*x + x - 6 = 0
First answer: 2
Second answer: -3
Correct!
```

19. CONCENTRATION. Whole numbers appear on the screen at regular intervals. As soon as a number appears for a second time, the user must hit the space bar. (This tests your reflexes as well as your memory.)

20. In the program State Your Terms a player must enter terms which are constructed correctly, according to BASIC. If they are incorrect the computer signals with SYNTAX ERROR. Change the program so that the computer does not exit from the program but diagnoses the mistake instead.

21. NEXT NUMBER, PLEASE. The computer begins a series of numbers and asks the player to supply the next in the series. Example:

```
1   6   11   16
Next number? 21
Correct!
```

42

The algorithm for this series is as follows:

$x_{n+1} = x_n + 5$ (recursive)   or

$x_n = 5*n-4$ (explicit).

You can program the following series:

    a) $x_n = a*n + b$

    b) $x_n = a*n + (-1)^n*b$

    c) $x_n = a*n^2 + b$

    d) $x_n = a*x_{n-1} + b$

    e) $x_n = a*x_{n-1} + b*x_{n-2} + c$

Caution!   A finite number series can be continued in an infinite number of ways.   But here we have to agree upon a correct and an incorrect answer, once the kind of series has been determined.

CHAPTER III

GAMES WITH GRAPHICS

Commercial video games make extensive use of graphics. Arcades are full of machines that produce fantastic creatures in imaginary landscapes; mountains that rise up and spew out invading armies; grids appear from nowhere and roads disappear at the bottom of the screen; planes and rockets fire across galaxies and objects explode into a thousand pieces and reform into new shapes. How does the computer produce this scene vibrating with color and activity? If we examine one of these game programs more closely to discover its secrets we will find graphic symbols, command characters and numbers, and the whole thing is bound to look pretty unintelligible.

Careful instructions are necessary in order to exploit the computer's capacity for graphic and acoustic programming. This chapter presents some basic techniques of direct memory access and teaches you how to use them with graphic and acoustic programming. You will also program games with graphics. The games have intentionally been kept very simple in order to focus on essentials. They represent the basic structures which you can later fill with content once you have mastered the techniques.

Example 3.1 STARTING WITH THE SCREEN
You are asked to program basic operations necessary for games using graphics. You will be drawing a frame and guiding a moving point across the screen.

A few things are important to know before tackling the problems below. A computer has countless memory cells, each one of which has the "address" of an integer. If you are working in "normal BASIC" you have no direct access to these

45

cells. But there are two commands that make direct memory operations possible: POKE and PEEK. The command

```
POKE <address>,<code number>
```

makes it possible to enter a number (between 0 and 255) directly into a memory cell with the given address. This number is the code for a command, a number value, or a character. For example, the command (POKE 59468,12) writes the number 12 into memory location 59468.

59468	12

   ↑   ↑
address   content

This content (namely, 12) of the designated memory cell is a command to the computer to shift to capital letters and graphic symbols. The function

```
PEEK (<address>)
```

is the content of the memory cell with the address named (in decimal form). PRINT PEEK(59468) prints "12" on the screen (after the above POKE command has been executed).

When you use these commands PEEK and POKE you get a pretty good idea of the process behind the computer's activity. PEEK allows us to have a look at a memory cell and POKE allows us to stick something into it.

NOTE: although POKE ... , ... is an independent command to the computer, PEEK(...) is the value of a function. Therefore, PEEK always has to be included in a command. For example, PRINT PEEK(...) or LET X = PEEK(...).

****

The video screen is made up of 40 * 25 = 1000 memory locations. Each of these contains the information about which character appears in what spot on the screen. The upper left corner of the screen corresponds to memory location number 32768; the lower right corner corresponds to the memory location with the address 33767. A table showing the screen may be useful:

46

Column							
		0	1	2	3		38	39
	0	32768	32769	32770	32771	...	32806	32807
	1	32808	32809	32810	32811	...	32846	32847
	2	32848	32849	32850	32851	...	32886	32887
	3	32888	32889	32890	32891	...	32926	32927
R o w		↑ Address of the memory location of the screen position row 3, column 2.						
	23	33688	33689	33690	33691	...	33726	33727
	24	33728	33729	33730	33731	...	33766	33767

If we want to enter a certain character into a predetermined spot on the screen, we need to give the POKE command with the address and code number of that character. For example, the command POKE 33267,42 puts the character * in the middle of the screen (33267 = 32768 + 499); the number 42 is the code number of the asterisk in the screen code. If we want to display the entire stock of characters on the screen we simply give the command:

```
FOR I = 0  TO 255 : POKE 32768 + I,I : NEXT I
```

****

Information on the screen code is available either from the user's manual of your computer or from the computer itself. For example, if we want the learn the code for the sign # then we type it in the upper left corner of the screen, move the cursor to the next line and write PRINT PEEK(32768). The number 35 will appear. This is the code number we are looking for. If we are trying the reverse and want to learn the character for a code number, then the screen will look as follows:

```
POKE 32848,35
#
READY
```

This little program makes things a bit easier:

47

```
00100 : PRINT CHR$(147)
00110 : PRINT "        USING THE SCREEN CODE
00120 : PRINT "        --------------------
00130 : PRINT
00140 : PRINT "        TO MOVE RIGHT: PRESS >
00150 : PRINT "        TO MOVE LEFT:  PRESS <
00160 : PRINT
00170 :
00180 : INPUT "CODE NUMBER OF CHARACTER"; C
00190 : IF C < 0 OR C > 255 THEN 180
00200 :
00210 : POKE 32768 + 10*40 + 9,C
00212 :
00215 : PRINT CHR$(19); : FOR G = 1 TO 9 : PRINT CHR$(17);
00220 : NEXT G : PRINT "CODE:    "; C
00222 :    REM HOME, AND DOWN 9
00223 :
00225 : PRINT CHR$(19); : FOR G = 1 TO 10 : PRINT CHR$(17);
00230 : NEXT G : PRINT "SYMBOL:
00235 :    REM HOME, AND DOWN 10
00240 :
00250 : GET T$ : IF T$ = ">" THEN LET C = C+1 : GOTO 210
00260 :          IF T$ = "<" THEN LET C = C-1 : GOTO 210
00270 :                                        GOTO 250
00280 : END
```

This creates the following screen display:

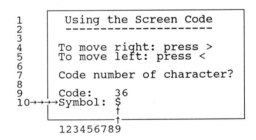

Line 210 gives the command to display the character with the
code number C in the slot 32768 + 10*40 + 9, i.e., in row
10, column 9, It is important not to confuse the screen
code with the ASCII code. Between 20 and 63 they are identical. But the ASCII code contains control characters in
addition to the displayable characters. However, it does
not contain the graphic characters. To get acquainted with
the ASCII code you only need to give the commands:

```
| PRINT ASC("$") |     or     | PRINT CHR$(36) |
```

Lines 215 through 230 of the above program for explor-
ing the screen code encourage us finally to describe the
functions involved in moving the cursor. There are CHR$
codes for:

1. Clearing the screen
2. Cursor to home slot (upper left corner)
3. Cursor to the right
4. Cursor to the left
5. Cursor downwards
6. Cursor upwards
7. Negative character display
8. Positive character display

Turn back to A WORD TO THE USER now, and review the
functions of the different CHR$ codes. Remember, though,
that each make of computer may have its own method of cursor
control.

At last we have the expertise to solve the problem
posed at the beginning of this chapter:
A. <u>Draw a frame around the edge of the screen.</u>
Solution: Begin with the upper horizontal bar. To draw it we
need to enter the code number Z of the chosen character into
the memory locations from A = 32768 through B = 32807:

```
| FOR I = A TO B STEP H : POKE I,Z : NEXT I |
```

and with a STEP-size of (H = 1). For the perpendicular bar
we need a spacing of H = 40 in order to skip over the lines.
This means A = 32847 and B = 33767.   We print the actual
character with a small subroutine and give it the parameters
A , B and H before calling it.   The selection of the character is accomplished by a multiple branch (line 290 in the
following program).

```
00100 : PRINT CHR$(147)
00110 : PRINT "       DRAWING A FRAME
00120 : PRINT "       ---------------
```

49

```
00140 REM DRAWS A FRAME AROUND SCREEN MARGIN
00150 :
00160 REM === CHOICE OF CHARACTERS ========================
00170 :
00180 : PRINT
00190 : PRINT "WHAT CHARACTER WILL YOU USE?
00200 : PRINT
00210 : PRINT "CROSS. . . . . . PRESS 1
00220 : PRINT "ASTERISK . . . . PRESS 2
00230 : PRINT "ARROW. . . . . . PRESS 3
00240 : PRINT "BALL . . . . . . PRESS 4
00250 : PRINT
00260 : GET A$ : IF A$ = "" THEN 260
00270 : IF A$ < "1" OR A$ > "4" THEN 260
00280 :
00290 : ON VAL(A$) GOTO 310, 320, 330, 340
00300 :
00310 : LET Z = 42 : GOTO 360 : REM ASTERISK *
00320 : LET Z = 43 : GOTO 360 : REM CROSS +
00330 : LET Z = 30 : GOTO 360 : REM ARROW ↑
00340 : LET Z = 81 :       360 : REM BALL  ○
00350 :
00360 : PRINT CHR$(147) : REM CLEAR THE SCREEN
00370 :
00380 REM === DRAWING THE TOP EDGE ========================
00390 :
00400 : LET A = 32768 : LET B = 32806
00410 : LET H = 1      : REM SPACE 1 (COULD BE OMITTED)
00420 : GOSUB 630      : REM TO SUBROUTINE 'FRAME CHARACTERS'
00430 :
00440 REM === DRAWING THE RIGHT PERPENDICULAR FRAME SIDE ===
00450 :
00460 : LET A = 32807 : LET B = 33727
00470 : LET H = 40     : REM SPACE 40
00480 : GOSUB 630      : REM TO SUBROUTINE 'FRAME CHARACTERS'
00490 :
00500 REM === DRAWING THE BOTTOM HORIZONTAL ================
00510 :
00520 : LET A = 33767 : LET B = 33729
00530 : LET H = -1     : SPACE -1
00540 : GOSUB 630      : REM TO SUBROUTINE 'FRAME CHARACTERS'
00550 :
00560 REM === DRAWING THE LEFT PERPENDICULAR FRAME SIDE ====
00570 :
00580 : LET A = 33728 : LET B = 32808
00590 : LET H = -40    : SPACE -40
00600 : GOSUB 630      : REM TO SUBROUTINE 'FRAME CHARACTERS'
00610 : END
00620 :
00630 REM +++ SUBROUTINE 'FRAME CHARACTERS' +++++++++++++++
00640 :
00650 : FOR I = A TO B STEP H : POKE I,Z : NEXT I : RETURN
```

This program for drawing a frame is long and detailed. Whenever it appears in future within another program we'll be able to keep it shorter. It is also possible to solve the problem using other methods, as shown by this example:

B. Move a point around the screen.

Its direction shall be determined by an appropriate spacing. Whenever it hits the frame it must jump back to the center of the screen. Here's how to do it: When you were drawing a frame on the screen, each character displayed was stationary. But with moving point graphics, the point must be e-rased each time it is to move in order to give the illusion of movement. With the command POKE I,81 the little ball O is placed in position I; with POKE I-1,32 the space char-acter (code number 32) is placed in position I-1 (where the ball had just been). This program makes a little ball run from left to right across the screen on line 3 (since we are considering the top line to be line 0):

```
10 : LET A = 32768 + 3*40
20 : LET B = A+39
30 : FOR I = A TO B
40 :    POKE I,81
50 :    POKE I-1,32
60 : NEXT I
```

It is important to become used to the various possibilities used to fix the direction of the moving point. In the fol-lowing program this factor is governed by the correct spac-ing H.

```
00100 : PRINT CHR$(147)
00110 : PRINT "     MOVING DOT
00120 : PRINT "     ----------
00130 :
00140 REM DEMONSTRATES THE PATH OF A MOVING CHARACTER
00150 REM DEPENDENT UPON SPACING
00160 :
00170 : LET P = 42        : REM CODE NUMBER OF THE CHARACTER
00180 : LET M = 32768 + 499  : REM CENTER OF SCREEN
00190 :
00200 : PRINT
```

51

```
00210 : INPUT "SPACING "; H
00220 :
00230 REM === DRAW FRAME =====================================
00240 :
00250 : LET A$ = "#######################################"
00260 : LET B$ = "#                                     #"
00270 : PRINT CHR$(147) A$;
00280 : FOR I = 1 TO 22 : PRINT B$; : NEXT I
00290 : PRINT A$ CHR$(19)
00300 :
00310 REM === MOVING THE DOT ==============================
00320 :
00330 : LET X = M             : REM CENTER OF SCREEN
00340 : POKE X,P              : REM POSITIONING THE DOT
00350 :
00360 : LET Y = X+H : REM NEW POSITION, LOOP START
00370 :
00380 : IF NOT(Y<32768 OR Y>33727 OR PEEK(Y) = 35) THEN 410
00385 : LET U = INT(960*RND(1)+40)
00390 : LET Y = 32768 + U : REM RANDOM POSITION
00400 :
00410 :    FOR I = 1 TO 100 : NEXT I      : REM WAITING LOOP
00420 :
00430 :    POKE Y,P      : REM POSITIONING THE DOT
00440 :    POKE X,32     : REM EXPUNGE OLD POSITION
00450 :    LET X = Y     : REM OLD POSITION BECOMES NEW ONE
00460 :
00470 : GOTO 360 : REM LOOP END
00480 :
00490 : END
```

Now for the third problem:

C. Use keyboard commands to guide a point cross the screen.

Solution: First draw a frame around the border of the screen. The frame defines the boundary of the moving point. Use the number keys

↖ 7	↑ 8	9 ↗
← 4	5	6 →
↙ 1	2 ↓	3 ↘

to guide the point, as shown in the illustration. Each direction corresponds to a different spacing. This table assigns every key a direction and thus a spacing. The following algorithm governs the control of the point:

```
MOVING DOT
x:= starting position
        Move dot to position
REPEAT
        Read key
        Determine the new position y
        IF y is on frame  THEN
          y :=x (* stay in old position *)
        END IF
        Move dot to position y
        Delete dot in old position x
        x:= y (* old position becomes the new
                   position y *)
END REPEAT
```

The command "Read key" is executed with a GET as follows: the number hit is entered into the variable T$. If no key is hit (T$ = ""), then the direction is unchanged; otherwise the new direction is determined by H(VAL(T$)). The query IF PEEK(Y) = R THEN LET Y = X prevents the dot from leaving the screen through the frame (instead, it remains in its previous spot).

```
00100 : PRINT CHR$(147)
00110 : PRINT "      MOVING DOT"
00120 : PRINT "      ----------"
00130 :
00140 REM A CHARACTER, RANDOMLY POSITIONED ON THE SCREEN,
00150 REM IS STEERED IN A DESIRED DIRECTION USING THE NUMBER
00160 REM KEYS. WHEN IT HITS THE FRAME, IT STOPS.
00170 :
00180 : LET R = 90  : REM CHARACTER FOR THE FRAME ("♦")
00190 : LET P = 81  : REM CHARACTER FOR THE DOT ("○")
00200 :
00210 REM === DRAWING THE FRAME ===========================
00220 :
00230 : PRINT CJR$(147)
00240 : FOR I = 32768 TO 32807
00250 :    POKE I,R : POKE 960+I,R
00260 : NEXT I
00270 : FOR I = 32768 TO 33727 STEP 40
00280 :    POKE I,R : POKE 39+I,R
00290 : NEXT I
00300 :
00310 REM === SETTING THE DIRECTION TABLE ==================
```

```
00320 :
00330 : DIM H(9)
00340 : LET H(1) =   39  :  REM LEFT AND DOWN
00350 : LET H(2) =   40  :  REM STRAIGHT DOWN
00360 : LET H(3) =   41  :  REM RIGHT AND DOWN
00370 : LET H(4) =   -1  :  REM LEFT
00380 : LET H(5) =    0  :  REM STATIONARY (STOP)
00390 : LET H(6) =    1  :  REM RIGHT
00400 : LET H(7) =  -41  :  REM LEFT AND UP
00410 : LET H(8) =  -40  :  REM STRAIGHT UP
00420 : LET H(9) =  -39  :  REM RIGHT AND UP
00430 :
00440 REM === FINDING RANDOM POSITION ====================
00450 :
00460 : LET X = INT(1000*RND(1)) + 32768
00470 : IF PEEK(X) = R THEN 460 : REM NOT AGAINST FRAME
00480 :
00490 REM === PLACING DOT AND STEERING ====================
00500 :
00510 : POKE X,P    : REM LOOP START, PLACING DOT -----------
00520 :
00530 :    GET T$                     : REM HIT KEY
00540 :    IF T$ <> "" THEN LET T = VAL(T$) : REM KEY WAS HIT
00550 :    LET Y = X + H(T)           : REM NEW POSITION
00560 :
00570 :    IF PEEK(Y) = R THEN LET Y = X : REM NOT AT FRAME
00580 :
00590 :    POKE Y,P                   : REM PLACING THE DOT
00600 :    POKE X,32                  : REM ERASE OLD POSITION
00610 :    LET X = Y                  : REM NEW BECOMES OLD
00620 :
00630 : GOTO 510   : REM LOOP END -----------------------
00640 :
00650 END
```

Before going much farther you should know a bit about
the "keyboard buffer."   A buffer is an "intermediate memory." A keyboard buffer is a type of memory that stores all
commands entered via the keyboard. An advantage of this
feature is that the user can continue typing characters into
the computer even though the machine is busy performing
other tasks. After these tasks are completed, the computer
can return to the keyboard buffer and process the information that has been entered in the interim. The following
little program demonstrates how to limit the number of characters in the keyboard buffer. This can be useful since it
gets troublesome if the computer continues to display keyboard characters after a game is over.

```
00100 : PRINT "      DEMONSTRATION OF THE KEYBOARD BUFFER
00110 : PRINT "      ------------------------------------
00120 : PRINT
00130 : PRINT "HOW MANY CHARACTERS WILL I RECALL (0 TO 9)";
00135 : INPUT N
00140 : IF N < 0 OR N > 9 THEN 120
00150 :
00160 : PRINT
00170 : PRINT "I WILL BE BUSY FOR A WHILE.
00180 : PRINT "ENTER A RANDOM NUMBER OF CHARACTERS.
00190 :
00200 : FOR I = 1 TO 5000 : NEXT I : REM SIMULATED ACTIVITY
00210 :
00220 : POKE 158,N : REM MAKE ROOM FOR N CHARACTERS
00230 :
00240 : PRINT
00250 : PRINT "I REMEMBERED THE ";N;
00255 : PRINT " CHARACTERS UNDER 'READY'"
00260 : PRINT "I AM NOW IN ENTER MODE
00270 :
00280 : END
```

Sometimes it is extremely useful to leave commands in the keyboard buffer for the computer to execute after it has finished a program (see, for example, the program 2.6, "State Your Terms"). If such a command is a cursor control code the computer will execute this by moving the cursor to the predetermined place on the screen. This is interpreted as a command when in direct mode (or as input in input mode). If a RETURN (code number 13) comes from the keyboard, then this acts as a RETURN in the direct mode (and in the input mode). This function makes it possible for a program to continue to work after it has been exited. It may be called a "programmed direct mode" (which is actually a contradiction).

Try the programming brainteasers 1 through 8 at the end of this chapter if you want to try out variations of the Moving Dot.

****

Now that we can steer a point across the screen at will, let's try a little game with graphics. The object is to place obstacles in the way of the point and thereby test the agility of the player who has to avoid them.

Example 3.2   SONAR SENSING (The Bat)

The player guides the point around objects in its way with the same uncanny power that guides a bat in its flawless night flying. The tricky part of this game is that the number of objects is governed by chance and keeps on increasing steadily.

The program has to have the following components:
1. Display of the game rules (if desired)
2. Initialization (determining the beginning values of the variables)
3. Drawing the "board", or field of play
4. Evaluation (of the game status or a player's success so far)
5. Exiting or continuing the game

The game itself (part 4) follows this algorithm:

```
Move point to starting position x
REPEAT
      Read key
      IF key not hit  THEN spacing :=1
      Determine new position y
      CANCEL IF  collision with object
      Set point in position y
      Delete point in position x
      x:= y (* old position becomes new *)
      Place new obstacle (* random *)
END REPEAT
```

Here is the program:

```
00100 : PRINT CHR$(147)
00110 : PRINT "     SONAR SENSING ('THE BAT')"
00120 : PRINT "     --------------------------"
00130 :
00140 : PRINT : PRINT
00150 : PRINT "          A SURVIVAL GAME"
00160 :
00170 REM A SIMPLE GAME OF REACTION AND SKILL
00180 :
00190 REM VARIABLES:
```

56

```
00200 :
00210 REM      P . . . . .   CODE OF DOT STEERED BY PLAYER
00220 REM      R . . . . .   CODE FOR FRAME
00230 REM      L . . . . .   CODE FOR EMPTY CHARACTER
00240 REM      N . . . . .   TIME CONSTANT
00250 REM      H(..) . . .   DIRECTION TABLE
00260 REM      X, Y  . . .   POSITIONS OF CHARACTER STEERED
00265 REM                    BY PLAYER
00270 REM      T$, T . . .   INPUT CHARACTERS (NUMBER KEYS)
00280 REM      H . . . . .   SPACING FOR DOT STEERING
00290 :
00300 :
00310 REM === RULES OF GAME ===============================
00320 :
00330 : PRINT : PRINT : PRINT : PRINT : PRINT
00340 : PRINT "SHALL I DISPLAY THE RULES? (Y/N)"
00350 : GET A$
00360 : IF A$ <> "Y" AND A$ <> "N" THEN 350
00370 : IF A$ = "N" THEN 480
00380 :
00390 : PRINT CHR$(147)
00400 : PRINT "     RULES"
00410 : PRINT "     -----"
00420 : PRINT : PRINT
00430 : PRINT "YOU CAN USE THE FOUR NUMBER KEYS 2,4,6,8  TO
00440 : PRINT "MOVE THE STAR IN THE NORMAL WAY. THE GAME
00450 : PRINT "ENDS IF YOU HIT AN OBSTACLE
00460 : PRINT "HIT A KEY"
00470 : GET Z$ : IF Z$ = "" THEN 470
00480 REM === INITIALIZATION ==============================
00490 :
00500 : LET H(2) = 40 : LET H(4) = -1 : LET H(6) = 1 :
00505 : LET H(8) = -40
00510 :
00520 : LET P = 42      : REM CODE NUMBER FOR STAR
00530 : LET R = 160     : REM CODE NUMBER OF ■
00540 : LET L = 32      : REM CODE NUMBER OF THE SPACE BAR
00550 : LET X = 33270   : REM STARTING POSITION
00560 : LET N = 150     : REM SETS SPEED OF DOT
00570 :
00580 REM === DISPLAY GAME FIELD===========================
00590 :
00600 : PRINT CHR$(147)
00610 :
00620 : FOR I = 32768 TO 32807
00630 :    POKE I,R : POKE I+960,R
00640 : NEXT I
00650 :
00660 : FOR I = 32768 TO 33727 STEP 40
00670 :    POKE I,R : POKE I+39,R
00680 : NEXT I
00690 :
```

```
00700 REM === GAME =========================================
00710 :
00720 : POKE X,P
00730 :
00740 : GET T$
00750 : IF T$ = "" THEN LET H = 1 : GOTO 780 : REM DOT TO
00755 :                                   REM RIGHT
00760 :    LET T = VAL(T$)
00770 :    LET H = H(T)
00780 :    LET Y = X+H
00790 :
00800 : IF PEEK(Y) <> L THEN 940          : REM COLLISION
00810 :
00820 :    POKE Y,P
00830 :    POKE X,L
00840 :    LET X = Y
00850 :
00860 : POKE INT(960*RND(TI))+32768,R    : REM OBSTACLE
00870 :
00880 : FOR I = 1 TO N : NEXT I : REM WAITING LOOP
00890 : LET N = 0.99*N     : REM ELAPSED TIME DECREASED
00900 :
00910 : GOTO 720
00920 :
00930 :
00940 REM === SCORE ========================================
00950 :
00960 : FOR I = 1 TO 150 : POKE X,R : POKE X,P : NEXT I
00970 :
00980 : PRINT CHR$(147)
00990 : PRINT : PRINT
01000 : PRINT "I AM AFRAID YOU LOSE."
01010 :
01020 :
01030 REM === EXIT OR REPEAT ===============================
01040 :
01050 : PRINT : PRINT
01060 : PRINT "ANOTHER TRY? (Y/N)"
01070 : GET A$
01080 : IF A$ <> "Y" AND A$ <> "N" THEN 1060
01090 : IF A$ = "Y" THEN RUN
01100 : PRINT
01110 : PRINT "GOOD-BYE."
01120 : PRINT : PRINT
01130 :
01140 : END
```

Explanation:

The dot is given its own momentum so the player cannot let it remain stationary. The spacing in line 750 is H = 1. This means that if no key is hit the point moves to the right. The time constant N (line 560) determines the speed of the dot by means of the waiting loop (line 880). The speed gradually increases because of the command in line 890: LET N = 0.99 * N. Line 960 causes the point to bounce and vibrate when it hits an obstruction: the characters * and ▮ are displayed in rapid succession in the same spot. Unfortunately, a player can't do anything except lose at this game. Try altering the program to give players better odds (see the brainteasers at the end of this chapter).

Lets help the frustrated players of Sonar Sensing get even with fate by giving them a chance. In this game they can blast those obstacles out of the way.

Example 3.3 STAR HUNT

The object is to devise a graphic game that functions like this: Within the framed screen there appear at random intervals a previously determined number of points (stars). The player hunts these down using a movable point (the spaceship). When the spaceship moves toward a star it can fire a shot in that direction which obliterates the star on impact. If the spaceship collides with a star it burns up and the game is over.

We divide the program into five subroutines:

```
01000 : PRINT CHR$(147)
01010 : PRINT "      STAR HUNT
01020 : PRINT "      ---------
01030 :
01040 REM TEST OF SKILL AND REACTION TIME
01050 :
01060 :
01070 REM *** MAIN PROGRAM *******************************
01080 :
01090 : GOSUB 1240 : REM DISPLAY RULES OF GAME
01100 :
01110 : GOSUB 1470 : REM GAME AREA AND STARS
01120 :
01130 : GOSUB 1870 : REM GAME
01140 :
01150 : GOSUB 2240 : REM SCORE
01160 :
01170 : GOSUB 2360 : REM EXIT OR REPEAT
01180 :
```

```
01190 : END
01200 :
01210 REM *** END OF MAIN PROGRAM *************************
01220 :
01230 :
01240 REM +++ SUBROUTINE DISPLAY OF GAME RULES ++++++++++++
01250 :
01260 : PRINT : PRINT
01270 : PRINT "DO YOU WANT TO SEE THE RULES? (Y/N)"
01280 : GET A$
01290 : IF A$ <> "Y" THEN IF A$ <> "N" THEN 1280
01300 : IF A$ = "N" THEN RETURN
01310 :
01320 : PRINT CHR$(147)
01330 : PRINT "      GAME RULES"
01340 : PRINT "      ----------
01350 : PRINT : PRINT
01360 : PRINT "YOUR SPACE SHIP LOOKS LIKE THIS:  O,"
01370 : PRINT "AND THESE ARE THE STARS YOU ARE HUNTING: *"
01380 : PRINT "YOU STEER THE SHIP WITH THE USUAL KEYS
01390 : PRINT "AND FIRE IN THE DIRECTION YOU ARE GOING.
01395 : PRINT "(WITH KEY NUMBER 5.)
01400 :
01410 : PRINT "EVERYTHING CLEAR? THEN PRESS '='"
01420 :
01430 : GET A$ : IF A$ <> "=" THEN 1410
01440 :
01450 : RETURN
01460 :
01470 REM ++ SUBROUTINE 'GAME AREA, STARS AND SPACE SHIP' ++
01480 :
01490 : PRINT : PRINT
01500 : PRINT CHR$(147);"HOW MANY STARS DO YOU WANT ";
01505 : INPUT N
01510 : IF N < 1 OR N > 50 OR INT(N) <> N THEN 1500
01520 :
01530 : REM --- DRAW FRAME
01540 :
01550 :    PRINT CHR$(147)
01560 :    FOR I = 32768 TO 32807
01570 :      POKE I,90 : POKE I+960,90
01580 :    NEXT I
01590 :
01600 :    FOR I = 32768 TO 33727 STEP 40
01610 :      POKE I,90 : POKE I+39,90
01620 :    NEXT I
01630 :
01640 : REM --- POSITION THE STARS
01650 :
01660 :    FOR I = 1 TO N
01670 :      LET S = INT(1000*RND(TI)) + 32768
01680 :      IF PEEK(S) = 42 OR PEEK(S) = 90 THEN 1690
```

```
01685 :     POKE S,42 : GOTO 1700
01690 :     LET I = I-1
01700 :   NEXT I
01710 :
01720 : REM --- POSITION THE SPACE SHIP
01730 :
01740 :   LET X = INT(1000*RND(TI)) + 32768
01750 :   IF PEEK(X) <> 32 THEN 1740
01760 :
01770 : REM === TABLE OF DIRECTIONS
01780 :
01790 :   DIM H(9)
01800 :   LET H(1)=39 : LET H(2)=40 : LET H(3)=41
01805 :   LET H(4)=-1 : LET H(6)= 1 : LET H(7)=-41
01810 :   LET H(8)=-40: LET H(9)=-39
01820 :
01830 : LET E = 0                         : REM COUNTS THE HITS
01840 :
01850 : RETURN
01860 :
01870 REM +++ SUBROUTINE GAME ++++++++++++++++++++++++++++++++
01880 :
01890 : POKE X,87              : REM SPACE SHIP APPEARS
01900 : FOR I = 1 TO 50 : NEXT I    : REM WAITING LOOP
01910 :
01920 : GET T$
01930 : IF T$ = "5" THEN 2030         : REM FIRE
01940 :   IF T$ <> "" THEN T = VAL(T$): REM KEY WAS PRESSED
01950 :     LET Y = X + H(T)          : REM NEW POSITION
01960 :
01970 :   IF PEEK(Y) = 42 THEN E$ = "BANG": RETURN
01975 :                   REM COLLISION
01980 :   IF PEEK(Y) = 90 THEN LET Y = X
01985 :                   REM TOUCHED FRAME
01990 :
02000 : POKE X,32              : REM ERASE OLD POSITION
02010 : LET X = Y : GOTO 1890      : REM JUMP BACK
02020 :
02030 : REM +++ FIRE +++
02040 :
02050 :   LET Z = X              : REM POSITION OF BULLET
02060 :   FOR I = 1 TO 50        : REM SHOT LOOP
02070 :     LET U = Z + H(T)     : REM NEW POSITION OF BULLET
02080 :     LET P = PEEK(U)
02090 :     IF P = 90 THEN 1920     : REM MISSED, NEW KEY
02100 :     POKE U,46               : REM DOT AS BULLET
02110 :     IF P = 42 THEN 2150     : REM A HIT
02120 :     POKE U,32 : LET Z = U   : REM ERASE AND REMEMBER
02125 :                             REM POSITION
02130 :   NEXT I
02140 :
02150 :   POKE U,86              : REM DETONATION
```

```
02160 :    FOR K = 1 TO 50 : NEXT K
02170 :    POKE U,32
02180 :    LET E = E+1               : REM HITS ADDED
02190 :    IF E < N THEN 1920        : REM JUMP BACK
02200 :    LET E$ = "SUCCESS"        : REM ALL STARS GONE
02210 :
02220 : RETURN
02230 :
02240 REM +++ SUBROUTINE 'SCORE' +++++++++++++++++++++++++++++
02250 :
02260 : PRINT CHR$(147)
02270 : IF E$ = "SUCCESS" THEN 2320
02280 : PRINT CHR$(18); :            REM REVERSE ON
02282 : FOR ZZ = 1 TO 10
02285 :    PRINT CHR$(29) CHR4(17);
02287 : NEXT ZZ :                    REM CURSOR LEFT AND DOWN
00290 : PRINT "COLLISION!" : GOTO 2330
02300 : RETURN
02310 :
02320 : PRINT "YOU DID IT!"
02330 : FOR I = 1 TO 1000 : NEXT I
02340 : RETURN
02350 :
02360 REM +++ SUBROUTINE 'EXIT OR REPEAT' +++++++++++++++++++
02370 :
02380 : PRINT CHR$(147)
02390 : PRINT : PRINT : PRINT
02400 : PRINT "WANT ANOTHER ROUND? (Y/N) "
02410 : GET A$ : IF A$<> "Y" THEN IF A$ <> "N" THEN 2410
02420 : IF A$ = "Y" THEN RUN
02430 : PRINT CHR$(147) "IF YOU CAN'T TAKE THE HEAT,";
02435 : PRINT " GET OUT OF THE KITCHEN!"
02440 : PRINT "BE SEEING YOU."
02450 :
02460 : RETURN
```

The most interesting aspect here is the subroutine "game."
When the spaceship is moved you have to know if it has hit a
star:

```
1970 : IF PEEK(Y) = 42 THEN E$ = "BANG" : RETURN
```

In this case the variable E$, which notes success or fail-
ure, is assigned the word "BANG." The subroutine then
ends. Firing a shot is not all that easy. The spaceship's
course is contained in the variable H(T). This direction is
communicated to the shot in line 2070: LET U = Z + H(T). If
a shot misses it hits the frame. Then PEEK(U) = 90. In
this case we skip to 1920 GET T$ and wait for a new key to

be hit that will move the spaceship along in some direction. A tiny explosion follows when a star is hit. The character "X" appears instantaneously. The entire shot is located in a counting loop to make this process clearer and more systematic: 2060 FOR I=1 TO 50. The program leaves this loop early when the shot hits either the wall or a star.

****

After all the shooting it's time to meet a peaceful animal.

Example 3.4 THE LIZARD'S SNACK
A lizard creeps across the screen.
Overhead buzzes a fly -- until the
lizard's tongue whips out and
devours the insect.

Up til now we have been
moving objects of one character across the screen.   For
this game we are going to construct our lizard out of several characters like this:

Do you see the resemblance?
These characters have the following code numbers: % (37), =
(61), - (64), . (46), ▲ (122),
< (60), > (62) and _ (121).
If we assign position K to the mouth part _ , then the front leg has the position K + 39. So we give the command POKE K + 39, 60. The entire lizard is drawn in lines 510 through 550 of the following program.

    The lizard's motion proceeds in a FOR...NEXT loop (lines 490 to 670). If K is raised by a step of 1, the lizard moves a step to the right and we have to fill with empty spaces those positions of the lizard's body that are not occupied by the characters (i.e., body parts) that get towed along.

While the lizard is creeping forward it can shoot its tongue upwards perpendicularly. This function is very similar to firing shots, as in the previous program. Here we have put it in a small subroutine. The lizard has 60 seconds to hunt for flies, then the game ends. The time is displayed at the top of the picture. The clock is set at 0 in line 330 and displayed in line 570.

```
00100 : PRINT CHR$(147)
00110 : PRINT "      THE LIZARD'S SNACK
00120 : PRINT "      ------------------
00130 :
00140 REM A GAME TO TEST SKILL AND REACTION TIME
00150 :
00160 REM === GAME RULES ================================
00170 :
00180 : PRINT
00190 : PRINT "THE LIZARD TRIES TO SNAP UP THE FLY.
00200 : PRINT "HIT ANY KEY TO ACTIVATE THE TONGUE.
00210 : PRINT
00220 : PRINT "NOW PRESS A KEY
00230 :
00240 : GET T$ : IF T$ = "" THEN 240
00250 :
00260 REM === PREPARATION ================================
00270 :
00280 : PRINT CHR$(147)
00285 : FOR ZZ = 1 TO 10 : PRINT CHR$(17); : NEXT ZZ
00290 :    REM DOWN 10
00295 : PRINT TAB(14) "GET READY!"
00300 :
00310 : FOR I = 1 TO 1000 : NEXT I : REM ADVANCE WARNING
00320 :
00330 : LET TI$ = "000000" : LET TO=TI : REM CLOCK SET
00340 :
00350 : LET Z = 0                      : REM FLY COUNTER
00360 :
00370 REM === GAME ========================================
00380 :
00390 : PRINT CHR$(147)
00400 :
00410 : REM --- POSITION THE FLIES
00420 :
00430 :    LET R = INT(10*RND(TI))
00440 :    LET X = 32768 + 15 + R*40 + INT(29*RND(TI))
00450 :    POKE X,42
00460 :
```

```
00470 : REM --- LET THE LIZARD RUN
00480 :
00490 :    FOR K = 33697 TO 33727 : REM K IS LIZARD'S HEAD
00500 :
00510 :       POKE K-1,32 : POKE K,32 : POKE K-41,32
00515 :       POKE K-40,32 : POKE K, 121 : POKE K-1,37
00520 :       POKE K-2,37 : POKE K-3,37 : POKE K-4,37
00530 :       POKE K-5,61 : POKE K-6,61 : POKE K-7,61
00540 :       POKE K-8,64 : POKE K-40,122 : POKE K-39,46
00545 :       POKE K-9,32 : POKE K+39,60 : POKE K+38,32
00550 :       POKE K+36,62 : POKE K+35,32
00560 :
00570 :       PRINT CHR$(19) "TIME: "; TI$ : REM NOTE TIME
00580 :
00590 :       GET A$
00600 :       IF A$ = "" THEN 670 : REM LIZARD'S NEXT STEP
00610 :
00620 :       GOSUB 710 : REM CATCH ATTEMPT
00630 :
00640 :       IF (TI-T0)/60 < 60 THEN 370 : REM A NEW FLY
00650 :       GOTO 1080                    : REM TIME RUNS OUT
00660 :
00670 :    NEXT K
00680 :
00690 :    GOTO 370                        : REM NEW ATTEMPT
00700 :
00710 REM +++ SUBROUTINE: CATCH ATTEMPT ++++++++++++++++++++
00720 :
00730 : REM --- TONGUE SHOOTS UPWARDS ---
00740 :
00750 :    FOR T = K-80 TO K-960 STEP -40
00760 :       POKE T,103
00770 :
00780 :       IF PEEK(T-40) <> 42 THEN 950 : REM NOTHING
00785 :                                    : REM CAUGHT
00790 :       FOR H = T TO K STEP 40       : REM FLY CAUGHT
00800 :          POKE H-40,32 : POKE H,81
00810 :          FOR I = 1 TO 5 : NEXT I
00820 :       NEXT H
00830 :
00840 :       REM --- DEVOUR SNACK ---
00850 :
00860 :       POKE K-39,32 : POKE K-40,32 : POKE K,98
00870 :       POKE K+1,98
00880 :       FOR I = 1 TO 50 : NEXT I
00890 :       POKE K,98 : POKE K+1,111
00900 :
00910 :       GOSUB 990 : RETURN : REM SWALLOW
00920 :
00930 :       POKE T,103 : POKE T,32
00940 :
00950 :    NEXT T
```

```
00960 :
00970 : RETURN
00980 :
00990 : REM +++ SUBROUTINE SWALLOW +++
01000 :
01010 :    PRINT : PRINT "SLURP!
01020 :    LET Z = Z+1
01030 :    PRINT : PRINT Z;" FLIES CAUGHT"
01040 :    FOR I = 1 TO 1000 : NEXT I : REM PAUSE TO DIGEST
01050 :
01060 : RETURN
01070 :
01080 REM === GAME END ======================================
01090 :
01100 : PRINT : PRINT : PRINT
01110 : PRINT "THE MINUTE IS UP.
01120 :
01130 : END
```
****

The following example is relatively complex. The object is to show how difficult it is to give a long program a clear and comprehensible structure. Sorry, there's more shooting in this one.

Example 3.5 BALLOONS
Balloons float to the ground out of the sky. The object is to zap them with a movable gun before they touch the ground.

This is how the screen looks:

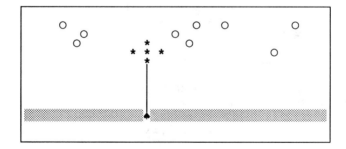

The program has the familiar general structure of "Introduction - Initial Values - Playing Field - Game -

66

Evaluation. The program section "Game" is the difficult one
because three different movements have to be coordinated:

    a) the back and forth movement of the gun

    b) the balloons floating downwards (moving at random)

    c) the shot

These parts are marked by comments, but the program is still
anything but transparent. Try to get some expert assist-
ance.

```
01000 : PRINT CHR$(147)
01010 : PRINT "      BALLOONS
01020 : PRINT "      --------
01030 :
01040 REM GRAPHIC GAME TESTING SKILL AND REACTION TIME
01050 :
01060 REM *** MAIN PROGRAM *******************************
01070 :
01080 : GOSUB 1220 : REM GAME DESCRIPTION
01090 :
01100 : GOSUB 1370  : REM BEGINNING VALUES
01110 :
01120 : GOSUB 1540  : REM GAME AREA
01130 :
01140 : GOSUB 1820  : REM GAME
01150 :
01160 : GOSUB 2620  : REM SCORE
01170 :
01180 : END
01190 :
01200 REM *** END MAIN PROGRAM ****************************
01210 :
01220 REM +++ SUBROUTINE GAME DESCRIPTION +++++++++++++++++++
01230 :
01240 : PRINT "YOU MUST SHOOT BALLOONS AS THEY DRIFT DOWN
01250 : PRINT "FROM THE SKY.  YOU GET 25 TRIES FOR 10
01260 : PRINT "BALLOONS. YOUR CANNON WILL KEEP MOVING BACK
01270 : PRINT "AND FORTH ACROSS THE GROUND.  YOU WILL SHOOT
01280 : PRINT "WITH THE " CHR$(18) "SPACE BAR" CHR$(146)
01290 : PRINT "HIT A KEY AND OFF WE GO!
01300 :
01310 : GET Q$ : IF Q$ = "" THEN 1310
01320 :
01330 : PRINT CHR$(147)
01340 :
01350 : RETURN
01360 :
01370 REM +++ SUBROUTINE BEGINNING VALUES +++++++++++++++++++
01380 :
01390 : ZK =    193  : REM CANNON
01400 : ZB =     81  : REM BALLOON
01410 : ZS =     93  : REM BULLET
```

```
01420 : ZR =    102  : REM FRAME
01430 : LE = 33528   : REM LEFT CORNER
01440 : RE = 33567   : REM RIGHT CORNER
01450 : BP = 33548   : REM BASE POINT (CANNON)
01460 : BR =      1  : REM DIRECTION
01470 : AB =     10  : REM NUMBER OF BALLOONS
01480 : SZ =      0  : REM NUMBER OF THE SHOT
01490 : VZ =     25  : REM NUMBER OF ATTEMPTS
01500 : ZL =     32  : REM SPACE
01510 :
01520 : RETURN
01530 :
01540 REM +++ SUBROUTINE GAME AREA ++++++++++++++++++++++++++
01550 :
01560 : FOR I = 32768 TO 32807
01570 :    POKE I,ZR
01580 :    POKE 760+I,ZR
01590 :    POKE 960+I,ZR
01600 : NEXT I
01610 :
01620 : FOR I = 33568 TO 33688 STEP 40
01630 :    POKE I,ZR
01640 :    POKE 39+I,ZR
01650 : NEXT I
01660 :
01670 : DIM B(40)
01680 :
01690 : FOR I = 1 TO 10
01700 :    LET K = INT(40*RND(TI)) + 32807
01710 :    IF PEEK(K) <> 32 THEN 1700
01720 :    LET B(I) = K
01730 :    POKE K,ZB
01740 : NEXT I
01750 :
01760 : POKE BP,ZK
01765 :
01770 : PRINT CHR$(19); : REM HOME
01775 : FOR ZZ = 1 TO 21 : PRINT CHR$(17); : NEXT ZZ
01780 :    REM DOWN TO BOTTOM
01785 : PRINT "ATTEMPTS       SHOTS       BALLOONS"
01790 :
01800 : RETURN
01810 :
01820 REM +++ SUBROUTINE GAME +++++++++++++++++++++++++++++++
01830 :
01835 : PRINT CHR$(19);
01840 : FOR ZZ = 1 TO 22 : PRINT CHR$(17); : NEXT ZZ
01845 : PRINT TAB(4) VZ CHR$(157)
01850 : PRINT TAB(18) CHR$(157) SZ TAB(31) AB CHR$(157)
01860 :
01870 : GET A$ : IF A$ = " " THEN 1960   : REM SHOT WITH
01875 :                                    REM SPACE BAR
```

68

```
01880 :
01890 : REM --- BACK-AND-FORTH MOVEMENT OF CANNON ---
01900 :
01910 :    IF NOT(LE < BP AND BP < RE) THEN LET BR = -BR
01920 :    POKE BP,ZR : LET BP = BP+BR : POKE BP,ZK
01930 :
01940 :    GOTO 1870 : REM HIT KEY AGAIN
01950 :
01960 : REM --- SHOT AND BALLOON FALL ---
01970 :
01980 :    LET VZ = VZ-1 : LET SZ = SZ+1 : LET P = BP
01990 :
02000 :    LET P = P-40
02010 :    IF PEEK(P) <> ZB THEN 2070      : REM HAVE NOT HIT
02015 :                                      REM BALLOON YET
02020 :    LET AB = AB-1 : REM A HIT!  ONE LESS BALLOON
02030 :    GOSUB 2320     : REM DETONATION
02040 :    GOSUB 2540     : REM ERASE ONE LOW-LEVEL BALLOON
02050 :    GOTO 2140      : REM ERASE SHOT
02060 :
02070 :    IF PEEK(P) <> ZR THEN 2110 : REM HAS NOT HIT FRAME
02080 :    POKE P+40,ZL             : REM SHOT HITS FRAME
02090 :    GOTO 2140                : REM FOR ERASING
02100 :
02110 :    POKE P,ZS                : REM BULLET
02120 :    GOTO 2000                : REM BULLET MOVES FORWARD
02130 :
02140 :    FOR I = BP-680 TO BP-40 STEP 40
02150 :      POKE I,ZL              : REM ERASE SHOT
02160 :    NEXT I
02170 :
02180 : IF AB <= 0 OR VZ <= 0 OR VZ < AB THEN RETURN
02185 :    REM GAME OVER
02190 :
02200 :    FOR I = 1 TO 10          : REM BALLOONS FALL
02210 :      LET K = INT(10*RND(TI))+1
02220 :      IF B(K) = 0 THEN 2270
02230 :      LET B(K) = B(K) + 40
02240 :      IF PEEK(B(K)) = ZR THEN LET VZ = 0
02245 :        REM LOW-LEVEL BALLOON
02250 :      POKE B(K) - 40, ZL
02260 :      POKE B(K),ZB
02270 :    NEXT I
02280 :
02290 :
02300 : GOTO 1840
02310 :
02320 : REM +++ SUBROUTINE 'DETONATION' +++
02330 :
02340 :    FOR I = P-2 TO P+2 : POKE I,42 : NEXT I
02350 :
02360 :    FOR I = P-80 TO P+80 STEP 40
```

69

```
02370 :     IF PEEK(I) = ZR THEN 2390
02380 :     POKE I,42
02390 :   NEXT I
02400 :
02410 :   FOR I = 1 TO 14
02420 :     IF B(I) = P THEN LET B(I) = 0 : GOTO 2450
02430 :   NEXT I
02440 :
02450 :   FOR I = P-2 TO P+2 : POKE I,32 : NEXT I
02460 :
02470 :   FOR I = P-80 TO P+80 STEP 40
02480 :     IF PEEK(I) = ZR THEN 2500
02490 :     POKE I,32
02500 :   NEXT I
02510 :
02520 : RETURN
02530 :
02540 : REM + SUBROUTINE 'DESTROYING A LOW-FLYING BALLOON' +
02550 :
02560 :   FOR I = 1 TO 14
02570 :     IF B(I) = P THEN B(I) = 0
02580 :   NEXT I
02590 :
02600 : RETURN
02610 :
02620 REM +++ SUBROUTINE 'SCORE' +++++++++++++++++++++++++++
02630 :
02640 : IF AB > 0 THEN LET Z$ = " LOST!" : GOTO 2660
02650 :               LET Z$ = " WON!"
02660 : PRINT CHR$(147) "YOU" Z$ : PRINT CHR$(17);
02670 : PRINT CHR$(17) "CARE FOR ANOTHER GAME? (Y/N)"
02680 : GET Q$ : IF Q$ = "Y" THEN RUN
02690 : IF Q$ <> "N" THEN 2680
02700 :
02710 : RETURN
```

Example 3.6   THE MAZE
The computer and the player compete to see which one can
successfully get through a maze first.

We construct the program in the familiar way using the rou-
tines: (1) Display the rules of the game (if desired); (2)
Initialization; (3) Maze construction; (4) Game (traversing
the maze); and (5) Evaluation. So that a new maze is displayed with each round, we have to include some chance ele-

ments. First we set up horizontal bars with randomly placed separations:

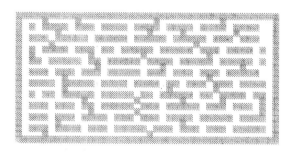

The width of these blocks is determined by the variable D1 in the following way:

```
1900 : LET ZU = INT(D1 * RND(TI) + 2)
```

Then random perpendicular bars are inserted whose thicknesses are determined by the value of the variable D2:

```
2010 : LET ZU = INT(D2 * RND(TI) + 5)
```

This gives us the method to construct the following sort of maze:

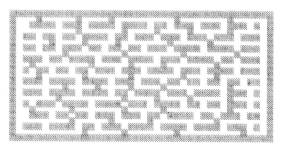

ΑΡΙΑΔΝΗ

The maze directly above (D1=3, D2=6) has no passage from the upper left (Start) to the lower right (Goal). To prevent this, we build a test run into our program that will test for this problem and, if necessary, break through the inner walls to create the necessary passage.

How does the computer find its way through the confusion of passageways and dead ends? It always runs down a

71

wall at the right edge. This means that it first tries a
step downwards and if this is blocked, it moves to the
right, then upwards, and last of all to the left. After
every step the computer takes, it is the player's turn to
steer a "mouse" through the maze by using the keys 2,4,6,8
in the way already familiar. If the player does not use the
opportunity to move, the computer goes ahead and takes another step.

The great advantage that the human player has over the
computer is the ability to scan the maze visually and tell
where the passages and the dead ends lie. The computer has
no such overview and has to explore every corner. The machine's only trump is its speed. It is precisely this dis-
tinction that is behind the "player versus computer" games
(such as chess). One of the most difficult tasks is to
teach the computer caution and the ability to evaluate complex situations. The following program is fairly extensive
even though it has been kept as simple as possible. Some
parts are still somewhat clumsily done (for example the
subroutine "The March through the Maze" or the way the player steers the mouse [lines 3490 through 3550]. So there is
plenty of room for your own improvement here. You will find
the method of cursor manipulation interesting. We use character sequences (in lines 1540 through 1590 and 2180, 2220,
2290, 2410, and 3630). Once we have defined a sequence of
control characters we can then pick as many from the sequence as we want.

```
01000 : PRINT CHR$(147)
01010 : PRINT "     MAZE
01020 : PRINT "     ----
01030 :
01040 REM GAME OF SKILL AND CONCENTRATION
01050 :
01060 :
01070 REM *** MAIN PROGRAM ********************************
01080 :
01090 : GOSUB 1230 : REM RULES OF PLAY
01100 :
01110 : GOSUB 1500  : REM INITIALIZATION
01120 :
01130 : GOSUB 1740  : REM MAZE CONSTRUCTION
01140 :
01150 : GOSUB 2340  : REM GAME
```

```
01160 :
01170 : GOSUB 3610  : REM SCORE AND EXIT
01180 :
01190 : END
01200 :
01210 REM *** END OF MAIN PROGRAM ************************
01220 :
01230 REM +++ SUBROUTINE 'RULES OF PLAY' +++++++++++++++++++
01240 :
01250 : PRINT
01260 : PRINT "SHALL I DISPLAY THE RULES? (Y/N)
01270 : GET A$ : IF A$ = "" THEN 1270
01280 : IF A$ <> "Y" THEN RETURN
01290 :
01300 : REM INSERT THE RULES AT THIS POINT.
01310 :
01480 : RETURN
01490 :
01500 REM +++ SUBROUTINE INITIALIZATION +++++++++++++++++++++
01510 :
01520 : REM CHARACTER SEQUENCES TO GOVERN CURSOR MOVEMENT
01530 :
01540 :    LET CD$ = CHR$(17) : REM CURSOR DOWN
01550 :    LET CR$ = CHR$(29) : REM CURSOR RIGHT
01560 :    FOR I = 1 TO 6
01570 :      LET CD$ = CD$ + CD$
01580 :      LET CR$ = CR$ + CR$
01590 :    NEXT I
01600 :
01610 : REM --- DETERMINING DEGREE OF DIFFICULTY ---
01620 :
01630 :    LET S = 20      : REM DETERMINES SPEED OF MOVEMENT
01640 :    LET D1 = 6      : REM DETERMINES DENSITY OF MAZE
01650 :    LET D2 = 12     : REM DETERMINES DENSITY OF MAZE
01660 :
01670 : REM --- START AND GOAL COORDINATES ---
01680 :
01690 :    LET SP = 32809 : REM STARTING POINT
01700 :    LET ZI = 33646 : REM GOAL
01710 :
01720 : RETURN
01730 :
01740 REM +++ SUBROUTINE 'MAZE CONSTRUCTION'  +++++++++++++++
01750 :
01760 : PRINT CHR$(147)
01770 :
01780 : REM --- HORIZONTAL OUTSIDE WALLS ---
01790 :
01800 :    FOR I = 32768 TO 32807 : POKE I,102 : NEXT I
01810 :    FOR I = 33648 TO 33687 : POKE I,102 : NEXT I
01820 :
01830 : REM --- HORIZONTAL INTERIOR WALLS ---
```

73

```
01840 :
01850 :    FOR P = 32848 TO 33568 STEP 80
01860 :      LET ZU = 0
01870 :      FOR I = 0 TO 39
01880 :        IF ZU > 0 THEN 1910
01890 :        IF ZU = 0 THEN LET ZU = -1 : GOTO 1930
01900 :        LET ZU = INT(D1*RND(TI)+2)
01910 :        POKE P+I,102
01920 :        LET ZU = ZU-1
01930 :      NEXT I
01940 :    NEXT P
01950 :
01960 : REM --- PERPENDICULAR CONNECTORS ---
01970 :
01980 :    LET ZU = 4
01990 :    FOR P = 32808 TO 33608 STEP 80
02000 :      FOR I = 0 TO 39
02010 :        IF ZU = 0 THEN POKE P+I,102
02015 :        IF ZU = 0 THEN LET ZU = INT(D2*RND(TI)+5)
02020 :        LET ZU = ZU-1
02030 :      NEXT I
02040 :    NEXT P
02050 :
02060 : REM --- PERPENDICULAR FRAME WALLS ---
02070 :
02080 :    FOR U = 32808 TO 33608 STEP 40
02090 :      POKE U,102
02100 :      POKE U+1,32
02110 :      POKE U+38,32
02120 :      POKE U+39,102
02130 :    NEXT U
02140 :
02150 : REM --- LABELLING ---
02160 :
02170 :    PRINT CHR$(19) CHR$(18) "START" CHR$(146)
02180 :    PRINT CHR$(19) LEFT$(CD$,22) LEFT$(CR$,36);
02185 :    PRINT CHR$(18) "GOAL" CHR$(146)
02190 :
02200 : REM --- TEST RUN ---
02210 :
02220 :    PRINT CHR$(19) LEFT$(CD$,23);
02230 :    PRINT "I AM LOOKING FOR A PATH
02240 :
02250 :    LET F$ = "TEST"
02260 :    GOSUB 2480  : REM MARCH THROUGH THE MAZE
02270 :    LET F$ = "SERIOUS"
02280 :
02290 :    PRINT CHR$(19) LEFT$(CD$,23);
02300 : PRINT "              A PATH EXISTS!
02310 :
02320 : RETURN
02330 :
```

```
02340 REM +++ SUBROUTINE 'GAME' ++++++++++++++++++++++++++++++
02350 :
02360 :   POKE SP,0 : REM COMPUTER'S MOUSE AT START POSITION
02370 :   POKE SP+1,38          : REM PLAYER'S MOUSE BESIDE IT
02380 :
02390 :   FOR I = 1 TO 1000 : NEXT I : REM WAITING LOOP
02400 :
02410 :   PRINT CHR$(19) LEFT$(CD$,23);
02420 :   PRINT "            OFF WE GO            ";
02430 :
02440 :   GOSUB 2480         : REM MARCH THROUGH THE MAZE
02450 :
02460 :   RETURN
02470 :
02480 REM +++ SUBROUTINE 'MARCH THROUGH THE MAZE' ++++++++++
02490 :
02500 :   LET PL = SP : LET PP = SP+1 : REM STARTING POSITION
02510 :   GOTO 2850              : REM FIRST, ONE STEP DOWNWARDS
02520 :
02530 : REM --- TRIAL STEPS ---
02540 :
02550 :   LET FE = PL+1     : REM STEP TO RIGHT
02560 :   IF PP = ZI THEN LET FE = PP : GOTO 3150
02565 :     REM END OF RUN
02570 :   IF FE = ZI THEN 3150
02580 :   IF PEEK(FE) <> 38 THEN 2620
02590 :   GOSUB 3310                    : REM JUMP OVER
02600 :   GOSUB 3430           : REM PLAYER CONTROLS 'MOUSE'
02610 :   IF U = 0 THEN 2550
02620 :   IF PEEK(FE) = 32 THEN GOSUB 3430 : GOTO 2850
02625 :     REM DOWNWARDS
02630 :
02650 :   LET FE = PL-40
02653 :   IF F$ = "SERIOUS" THEN 2660
02657 :   IF FE <= SP-40 THEN GOSUB 3180 : GOTO 2480
02660 :   IF PP = ZI THEN LET FE = PP : GOTO 3150
02680 :   IF PEEK(FE) <> 38 THEN 2720
02690 :   GOSUB 3310
02700 :   GOSUB 3430
02710 :   IF U = 0 THEN 2650
02720 :   IF PEEK(FE) = 32 THEN GOSUB 3430 : GOTO 2950
02725 :     REM UPWARDS
02730 :
02750 :   LET FE = PL-1      : REM STEP LEFT
02760 :   IF PP = ZI THEN LET FE = PP : GOTO 3110
02780 :   IF PEEK(FE) <> 38 THEN 2820
02790 :   GOSUB 3310
02800 :   GOSUB 3430
02810 :   IF U = 0 THEN 2750
02820 :   IF PEEK(FE) = 32 THEN GOSUB 3430 : GOTO 2650
02825 :     REM UPWARDS
02830 :
```

```
02850 :    LET FE = PL+40      : REM STEP DOWNWARDS
02860 :    IF PP = ZI THEN LET FE = PP : GOTO 3150
02870 :    IF FE = ZI THEN 3150
02880 :    IF PEEK(FE) <> 38 THEN 2920
02890 :    GOSUB 3310
02900 :    GOSUB 3430
02910 :    IF U = 0 THEN 2850
02920 :    IF PEEK(FE) = 32 THEN GOSUB 3430 : GOTO 2750
02925 :      REM GO LEFT
02930 :    GOTO 2550
02940 :
02950 :    LET FE = PL+1       : REM STEP RIGHT
02960 :    IF PP = ZI THEN LET FE = PP : GOTO 3150
02970 :    IF FE = ZI THEN 3150
02980 :    IF PEEK(FE) <> 38 THEN 3020
02990 :    GOSUB 3310
03000 :    GOSUB 3430
03010 :    IF U = 0 THEN 2950
03020 :    IF PEEK(FE) = 32 THEN GOSUB 3430 : GOTO 3050
03025 :      REM DOWNWARDS
03030 :    GOTO 2650
03040 :
03050 :    LET FE = PL+40       : REM STEP DOWNWARDS
03060 :    IF PP = ZI THEN LET FE = PP : GOTO 3150
03070 :    IF FE = ZI THEN 3150
03080 :    IF PEEK(FE) <> 38 THEN 3120
03090 :    GOSUB 3310
03100 :    GOSUB 3430
03110 :    IF U = 0 THEN 3050
03120 :    IF PEEK(FE) = 32 THEN GOSUB 3430 : GOTO 2750
03125 :      REM LEFT
03130 :    GOTO 2950
03140 :
03150 : REM --- END OF RUN ---
03152 :
03154 :    IF PP = FE THEN RETURN : REM END OF RUN
03156 :    GOSUB 3430
03158 :
03160 : RETURN
03170 :
03180 : REM ++++++++ SUBROUTINE 'BREAK THROUGH' ++++++++++
03190 :
03200 :    FOR K = 1 TO 10
03210 :      LET D = INT(840*RND(1))
03220 :      LET G = 32768 + D
03230 :      IF D < 41 OR D = 40*INT(D/40) THEN 3210
03240 :      IF ZI < G OR (D+1) = 40*INT((D+1)/40) THEN 3210
03250 :      IF PEEK(G) = 32 THEN 3210
03260 :      POKE G,32 : REM REMOVE OBSTACLE
03270 :    NEXT K
03280 :
03290 : RETURN
```

```
03300 :
03310 : REM +++ SUBROUTINE 'JUMP OVER' +++
03320 :
03330 :    LET U = 0
03340 :    IF PEEK(PP+1) <> 32 AND PEEK(PP-1) <> 32 THEN 3360
03350 :    GOTO 3390
03360 :    IF PEEK(PP+40)<>32 AND PEEK(PP-40) <> 32 THEN 3380
03370 :    GOTO 3390
03380 :    LET U = 1
03390 :    LET FE = PL
03400 :
03410 : RETURN
03420 :
03430 : REM +++ 'SWITCHING AND PLAYER CONTROL' +++
03440 :
03450 :    POKE PL,32       : REM ERASE OLD POSITION
03460 :    LET PL = FE      : REM NEW POSITION BECOMES OLD
03470 :    IF F$ = "TEST" THEN RETURN  : REM TEST RUN
03480 :    POKE PL,0            : REM SHOW COMPUTER'S MOUSE
03490 :    GET P$ : IF P$ <> "" THEN LET R$ = P$
03495 :       REM KEY PRESSED
03500 :    IF R$ = "2" THEN LET FE = PP+40 : GOTO 3550
03510 :    IF R$ = "4" THEN LET FE = PP-1  : GOTO 3550
03520 :    IF R$ = "5" THEN RETURN
03530 :    IF R$ = "6" THEN LET FE = PP+1  : GOTO 3550
03540 :    IF R$ = "8" THEN LET FE = PP-40 : GOTO 3550
03550 :    IF PEEK(FE) <> 32 THEN 3570
03555 :    POKE PP,32 : LET PP = FE : POKE PP,38
03560 :
03570 :    IF F$ = "SERIOUS" THEN FOR T = 1 TO S : NEXT T
03580 :
03590 : RETURN
03600 :
03610 REM +++ SUBROUTINE 'SCORE' +++++++++++++++++++++++++++
03620 :
03630 : PRINT CHR$(19) LEFT$(CD$,23);
03635 :    REM CURSOR AT BOTTOM LINE
03640 : IF PP = ZI THEN PRINT "   YOU HAVE ";: GOTO 3660
03650 : PRINT "    I HAVE ";
03660 : PRINT "WON THE GAME!    ";
03670 :
03680 : FOR Y = 1 TO 1000 : GET A$ : NEXT Y
03690 :
03700 : PRINT CHR$(147) : PRINT : PRINT
03710 : PRINT "ANOTHER ROUND? (Y/N)
03720 : GET A$ : IF A$ = "" THEN 3720
03730 : IF A$ = "Y" THEN RUN
03740 :
03750 : RETURN
03760 :
```

****

Sam Lloyd, America's most famous puzzle specialist and game inventor, published "Back from the Klondike" in 1907. Here is the problem:

Example 3.7   BACK FROM THE KLONDIKE

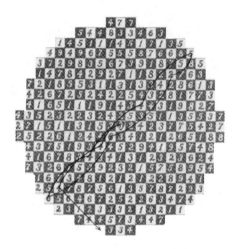

Begin at the heart in the middle. On your first day your journey takes you three paces in the direction of any of the eight compass points (North, South, East, West, or diagonally Northeast, Northwest, Southeast, Southwest). You then reach a square with a number. This tells you how many steps you may take on the second day of your journey. From each new square you march in any direction as many spaces as are designated in the square. You repeat this until you come to a square with a number that makes it possible for you to leave the forest with one move. Once you are out of the woods you've solved the puzzle.

Instead of drawing the maze on paper with a pencil, we'll draw it on the screen and enter the direction of our

march with key commands. The computer will evaluate the
path covered and will announce success or failure.

```
01000 : PRINT CHR$(147)
01010 : PRINT "     BACK FROM THE KLONDIKE
01020 : PRINT "     ----------------------
01030 :
01040 REM GAME BASED ON IDEA OF SAM LLOYD
01050 :
01060 :
01070 REM *** MAIN PROGRAM ******************************
01080 :
01090 : GOSUB 1240 : REM GAME DESCRIPTION
01100 :
01110 : GOSUB 1570  : REM INITIALIZATION
01120 :
01130 : GOSUB 2100  : REM FIELD OF PLAY
01140 :
01150 : GOSUB 1770  : REM GAME
01160 :
01170 : GOSUB 2570  : REM EXIT OR REPEAT
01180 :
01190 : END
01200 :
01210 REM *** END OF MAIN PROGRAM *************************
01220 :
01230 :
01240 REM +++ SUBROUTINE 'START' +++++++++++++++++++++++++++
01250 :
01260 : PRINT
01270 : PRINT "YOU ARE IN THE DIFFICULT POSITION OF HAVING
01280 : PRINT "TO FIND YOUR WAY OUT OF A LARGE FOREST. YOU
01290 : PRINT "CAN DETERMINE YOUR DIRECTION BY PUSHING KEYS
01300 : PRINT "1-9 (WITHOUT USING 5). BUT YOU ARE NOT OUT
01310 : PRINT "OF THE WOODS UNTIL YOU HAVE EXITED BY EXACTLY
01315 : PRINT "ONE SPACE. PRESS ANY KEY TO CONTINUE
01320 :
01330 : GET A$ : IF A$ = "" THEN 1330
01340 :
01350 : PRINT CHR$(147) : PRINT "THE REVERSE FIELD SHOWS
01360 : PRINT "YOUR POSITION. THE NUMBER SHOWS HOW MANY
01370 : PRINT "STEPS TO TAKE....LOTS OF LUCK!
01375 : PRINT "PRESS ANY KEY
01380 :
01390 : GET A$ : IF A$ = "" THEN 1390
01400 :
01410 : PRINT CHR$(147)
01420 : PRINT "THESE COMPASS POINTS ARE YOUR DIRECTIONS:
01430 : PRINT CHR$(17) "  8  :    NORTH
01440 : PRINT CHR$(17) "  9  :    NORTHEAST
01450 : PRINT CHR$(17) "  6  :    EAST
```

```
01460 : PRINT CHR$(17) "   3   :   SOUTHEAST
01470 : PRINT CHR$(17) "   2   :   SOUTH
01480 : PRINT CHR$(17) "   1   :   SOUTHWEST
01490 : PRINT CHR$(17) "   4   :   WEST
01500 : PRINT CHR$(17) "   7   :   NORTHWEST
01510 : PRINT CHR$(17) "   KEY '5' LETS YOU GIVE UP
01515 : PRINT "PRESS ANY KEY
01520 :
01530 : GET A$ : IF A$ = "" THEN 1530
01540 :
01550 : RETURN
01560 :
01570 REM +++ SUBROUTINE 'INITIALIZATION' +++++++++++++++++++
01580 :
01590 :    LET  P = 33260 : REM STARTING POSITION
01600 :    LET  D =       0 : REM NEXT DISTANCE
01610 :    LET WD =       0 : REM WHOLE DISTANCE
01615 :    LET DS =       0 : REM NUMBER OF DAYS
01620 :
01625 :    GOSUB 3000 : REM CURSOR CONTROL STRING
01630 :
01635 :
01640 : REM --- DIRECTIONS ---
01650 :
01660 :    FOR I = 1 TO 9
01670 :       READ B(I) : READ B$(I)
01680 :    NEXT I
01690 :    RESTORE
01700 :
01710 DATA 39,SOUTHWEST,40,SOUTH,41,SOUTHEAST
01720 DATA -1,WEST,,,1,EAST,-41,NORTHWEST
01730 DATA -40,NORTH,-39,NORTHEAST
01740 :
01750 : RETURN
01760 :
01770 REM +++ SUBROUTINE 'GAME' ++++++++++++++++++++++++++++++
01780 :
01790 : LET   D = PEEK(P)-176
01800 : LET  WD = WD + D
01810 : LET  DS = DS + 1
01820 :
01830 : POKE 32958,63
01840 : FOR I = 32959 TO 32967 : POKE I,32 : NEXT I
01850 :
01860 : GET R : IF R = 0 THEN 1860
01870 : IF B(R) = 0 THEN GOSUB 2760 : END
01875 :                                  REM PLAYER GIVES UP
01880 : IF PEEK(P+B(R)*D) <> 102 THEN 1895 : REM NOT OUT
01890 : IF PEEK(P+B(R)*(D-1)) <>102 THEN RETURN
01895 : IF PEEK(P+B(R)*D) < 49 THEN 1860
01900 : IF PEEK(P+B(R)*D) > 57 THEN 1860
01910 :
```

```
01920 : GOSUB 2010 : REM DISPLAY FIELD OF PLAY
01930 :
01940 : POKE P,D+48  : REM REVERT
01950 : LET P = P + (B(R)*D)
01960 : POKE P,(PEEK(P)+128)
01970 :
01980 : FOR I = 1 TO 1000 : NEXT I
01990 : GOTO 1770
02000 :
02010 : REM --- PRESS VARIABLES ---
02020 :
02025 : PRINT CHR$(19);
02030 : PRINT LEFT$(CD$,4); SPC(30); B$(R)
02040 : PRINT LEFT$(CD$,5); SPC(28); D
02050 : PRINT LEFT$(CD$,5); SPC(30-LEN(STR$(WD))); WD
02060 : PRINT LEFT$(CD$,5); SPC(32-LEN(STR$(DS))); DS
02070 :
02080 : RETURN
02090 :
02100 REM +++ SUBROUTINE 'DISPLAY' +++++++++++++++++++++++++
02110 :
02120 : PRINT CHR$(147)
02130 : PRINT "            "
02140 : PRINT "        477     "
02150 : PRINT "      544833463    "
02160 : PRINT "     1451114517135   "
02170 : PRINT "    494967555876685  "
02180 : PRINT "    37298356739187685 "
02190 : PRINT "    14784292711822763  "
02200 : PRINT "    7218553113133428613 "
02210 : PRINT "    4267252422543281773 "
02220 : PRINT "    4165111914344319827  "
02230 : PRINT "    435232232425351135537 "
02240 : PRINT "    271511315332423775427 "
02250 : PRINT "    25228124463412126518 "
02260 : PRINT "    437519344529419574 8 "
02270 : PRINT "    4167834341312323624 "
02280 : PRINT "    7326153923215758954 "
02290 : PRINT "    16734812121228941  "
02300 : PRINT "    25478756135787293 "
02310 : PRINT "    656467252263474   "
02320 : PRINT "     2312333213211   "
02330 : PRINT "      744573447    "
02340 : PRINT "        334     "
02350 : PRINT "            "
02360 :
02365 : REM POKE CODE FOR ' ' IS 102 (SEE LINE 1880)
02370 : POKE 33260,179 : REM INVERSE THREE
02380 :
02385 : PRINT CHR$(19);
02390 : PRINT LEFT$(CD$,2); TAB(27); "DIRECTION:"
02400 : PRINT TAB(27); "----------"
```

81

```
02410 :
02420 : PRINT LEFT$(CD$,3); TAB(27); "PATH:"
02430 : PRINT TAB(27); "-----"
02440 : PRINT TAB(32); "MILES"
02450 :
02460 : PRINT LEFT$(CD$,2); TAB(27); "WHOLE"
02470 : PRINT TAB(27); "DISTANCE:"
02480 : PRINT TAB(27); "---------"
02490 : PRINT TAB(32); "MILES"
02500 :
02510 : PRINT LEFT$(CD$,3); TAB(27); "DAYS:"
02520 : PRINT TAB(27); "----"
02530 : PRINT CHR$(19)
02540 :
02550 : RETURN
02560 :
02570 REM +++ SUBROUTINE 'EXIT OR REPEAT' ++++++++++++++++++
02580 :
02590 : PRINT CHR$(147)
02600 : PRINT CHR$(17) "      CONGRATULATIONS!
02610 : PRINT            "      ----------------
02620 : PRINT LEFT$(CD$,3)
02630 : PRINT "DESPITE EXPECTATIONS TO THE CONTRARY,
02640 : PRINT "YOU HAVE GOTTEN OUT OF THE WOODS!
02650 : PRINT "IT TOOK YOU"; DS; "DAYS.
02660 : PRINT "YOU COVERED A DISTANCE OF"; WD; "MILES.
02670 : PRINT "CARE TO TRY IT AGAIN? (Y/N)
02680 :
02690 : GET A$ : IF A$ = "" THEN 2690
02700 : IF A$ = "Y" THEN RUN
02710 : IF A$ <> "N" THEN 2670
02720 : PRINT CHR$(147)
02730 :
02740 : RETURN
02750 :
02760 REM +++ SUBROUTINE 'OBJECT OF THE GAME' +++++++++++++
02770 :
02775 : PRINT CHR$(147); LEFT$(CD$,3)
02780 : PRINT "TSK, TSK. READY TO QUIT ALREADY??
02790 : PRINT "FROM THE WAY YOU WERE PLAYING,
02800 : PRINT "I ADMIT I AM NOT SURPRISED.
02810 : PRINT "ALTHOUGH YOU'VE BEEN WANDERING
02820 : PRINT "AROUND IN THE FOREST FOR"; DS; "DAYS
02830 : PRINT "AND HAVE TRAVELLED"; WD; "MILES
02840 : PRINT "YOU ARE STILL FAR FROM YOUR GOAL.
02850 : PRINT "I KNOW YOU'LL DO BETTER NEXT TIME!
02860 :
02870 : RETURN
02880 :
03000 REM === SUB-SUBROUTINE CURSOR CONTROL STRING ===
03010 :
03020 :    LET CD$ = CHR$(17)
```

```
03030 :    FOR I = 1 TO 6
03040 :        CD$ = CD$ + CD$
03050 :    NEXT I
03060 :
03070 : RETURN
03080 :
```

****

Even though graphics games do not typically involve
one's sense of hearing, we are going to continue this chapter with a program that does use sound effects.

There are particular moments when the computer signals
the user with a beep. You may hear it when you turn on the
computer or when you've made a command error. How does this
work? Every microprocessor -- the "heart" of the computer
-- contains a device called the synchronizer that uses a
specific "heartbeat" to govern the rate of the switching
process. As we know from elementary physics, a tone or
sound is generated when the cone of a loudspeaker vibrates
at a certain frequency. Some microcomputers have a tiny
loudspeaker built in (or one can be added). All we have to
do is alter the frequency of the tone it produces.
Commodore PET computers, for example, have the option of a
shift register

83

0	1	1	0	1	0	1	1

which can push a bit pattern in a circle. Whether or not
this happens is determined by the memory cell with the address 59467. If this contains 16, then the shift register
may run freely, but usually it contains a zero (0). The
contents of memory cell number 59464 determines the frequency of revolutions. With every revolution a signal is given
to the built-in loudspeaker at the so-called user port.
Finally, the contents of cell number 59466 determine the
bit pattern in the shift register which governs the sound
produced. (This may seem a bit vague but you don't need any
more details at this point.) The following commands produce
tones:

POKE 59467,16   Tone on
POKE 59467,0   Tone off
POKE 59464,h (0 ≤ h ≤ 255)   pitch
POKE 59466,k (0 ≤ k ≤ 255)   timbre

Explore these possibilities using the following program:

```
00100 : PRINT CHR$(147)
00110 : PRINT "          MELODY MAKER
00120 : PRINT "          ------------
00130 :
00140 REM DEMONSTRATION PROGRAM FOR PRODUCING MUSICAL TONES.
00150 :
00160 : PRINT
00170 : PRINT "CONTINUOUS TONE . . . . . .   1
00180 : PRINT "SIREN . . . . . . . . . . .   2
00190 : PRINT "RANDOM MUSIC. . . . . . . .   3
00200 : PRINT "END . . . . . . . . . . . .   9
00210 :
00220 : GET A$ : IF A$ = "" THEN 220
00230 : ON VAL(A$) GOSUB 320, 630, 820
00240 :
00250 : IF A$ <> "9" THEN 160
00260 :
00270 : POKE 59464,0
00280 : POKE 59466,0
00290 : POKE 59467,0
00300 : END
00310 :
00320 REM +++ SUBROUTINE 'CONTINUOUS TONE' +++++++++++++++++
00330 :
00340 : PRINT CHR$(147)
00350 : PRINT "YOU CAN PRODUCE A CONTINUOUS TONE WITH PITCH
```

```
00360 : PRINT "AND TIMBRE OF YOUR CHOICE.
00370 : PRINT
00380 : INPUT "PITCH (BETWEEN 1 AND 255)      "; H
00390 : IF H < 1 OR H > 255 THEN 380
00400 : INPUT "TIMBRE (BETWEEN 1 AND 255)      "; K
00410 : IF K < 1 OR K > 255 THEN 400
00420 :
00430 : PRINT
00440 : PRINT "TONE ON --> PRESS A KEY
00450 : GET A$ : IF A$ = "" THEN 450
00460 :
00470 : POKE 59467,16  : REM TURN ON SYNCHRONIZER
00480 : POKE 59464,H   : REM CHOOSE PITCH
00490 : POKE 59466,K   : REM CHOOSE TIMBRE
00500 :
00510 : PRINT
00520 : PRINT "IF YOU WOULD LIKE A DIFFERENT PITCH OR
00530 : PRINT "TIMBRE, PRESS 'A'.
00540 : PRINT "IF YOU WANT TO RETURN TO THE MENU, PRESS '@'.
00550 :
00560 : GET A$ : IF A$ = "" THEN 560
00570 :
00580 : IF A$ = "A" THEN 370
00590 : IF A$ = "@" THEN POKE 59467,0 : RUN
00600 :
00610 : GOTO 510
00620 :
00630 REM +++ SUBROUTINE 'SIREN' +++++++++++++++++++++++++++
00640 :
00650 : PRINT CHR$(147)
00660 : PRINT "         SIREN
00670 : PRINT "         -----
00680 :
00685 : PRINT
00690 : PRINT "TO END, PRESS ANY KEY.
00700 :
00710 : POKE 59467,16                     : REM TONE ON
00720 : POKE 59466,1                      : REM TIMBRE
00730 : FOR I = 10 TO 245 : POKE 59464,I : NEXT I
00735 :    REM INCREASE PITCH
00740 : FOR I = 245 TO 10 STEP -1 : POKE 59464,I : NEXT I
00750 :
00760 : GET A$ : IF A$ = "" THEN 710
00770 :
00780 : POKE 59467,0                      : REM TONE OFF
00790 :
00800 : RETURN
00810 :
00820 REM +++ SUBROUTINE 'RANDOM MUSIC' ++++++++++++++++++++
00830 :
00840 : PRINT CHR$(147)
00850 : PRINT "         RANDOM MUSIC
```

```
00860 : PRINT "           ------------
00870 :
00875 : PRINT
00880 : PRINT "TO END, PRESS ANY KEY.
00890 :
00900 : POKE 59467,16                : REM TONE ON
00910 : POKE 59466,50
00920 : LET Z = INT(255*RND(TI)+1)   : REM RANDOM FREQUENCY
00930 : POKE 59464,Z                 : REM RANDOM PITCH
00940 :
00950 : GET A$ : IF A$ = "" THEN 920
00960 :
00970 : POKE 59467,0                 : REM TONE OFF
00980 :
00990 : RETURN
01000 :
                        **************
```

Example 3.8   MELODY MAKER

The object is to write a program that has the computer play pieces of music.

Here is the "Badinerie" from Bach's B minor suite:

A piece of music consists of a series of pairs: the note and its length. The length is the duration of the tone indicated by the note. This is the form used to write our musical composition in DATA lines:

```
A,2,Z,2,C1,2,A,2,...
```

These pairs are entered in the variable pair R$,T. The transformation from R$ into the appropriate frequency-determining number takes place in a subroutine, i.e.

```
510 : IF R$ = "A" THEN LET R = 140 : RETURN
```

This means that R = 140 is entered into memory cell number 59464 to create the tone A. The length is produced by the waiting loop

```
FOR I = 1 TO T1*T - T2 : NEXT I
```

in which T1 and T2 are the appropriate constants. The algorithm looks like this:

MUSICAL COMPOSITION

```
REPEAT
    Read note R$ and length T
    Translate R$ into the frequency-determining number R
    Write R into memory cell 59464
    Run through waiting loop using duration T
UNTIL  R$ = "End"
```

Here is the program:

```
00100 : PRINT CHR$(147)
00110 : PRINT "          BADINERIE
00120 : PRINT "          ---------
00130 :
00140 REM PROGRAM TO DEMONSTRATE MUSICAL COMPOSITION
00150 :
00155 : PRINT : PRINT : PRINT
00160 : PRINT "     DO YOU LIKE BACH?" : PRINT
00170 : PRINT " THEN YOU'LL LIKE THE" : PRINT
00180 : PRINT "   BADINERIE FROM THE B MINOR SUITE" : PRINT
00190 : PRINT "      FOR FLUTE AND ORCHESTRA." : PRINT
00200 : PRINT "              PRESS A KEY.
00210 :
00220 : GET A$ : IF A$ = "" THEN 220
00230 :
```

```
00240 REM *** MAIN PROGRAM ********************************
00250 :
00260 : LET T1 = 5              : REM SETS THE TEMPO
00270 :
00280 : POKE 59467,16          : REM TONE ON
00290 : POKE 59466,10          : REM TIMBRE
00300 : POKE 59464,0           : REM FREQUENCY ZERO
00310 :
00320 : READ R$,T              : REM ENTER NOTE AND LENGTH
00330 :   IF R$ = "@" THEN 400 : REM PIECE IS OVER
00340 :   GOSUB 470     : REM TRANSITION NOTE -> FREQUENCY
00350 :   POKE 59464,0          : REM FREQUENCY ZERO
00360 :   POKE 59464,R          : REM FREQUENCY R
00370 :   FOR I = 1 TO T*T1 - 100 : NEXT I : REM LENGTH T
00380 : GOTO 320
00390 :
00400 : POKE 59464,0           : REM FREQUENCY ZERO
00410 : POKE 59466,0           : REM TIMBRE ZERO
00420 : POKE 59467,0           : REM TONE OFF
00430 :
00440 : END : REM *** END OF MAIN PROGRAM *****************
00450 :
00460 :
00470 REM +++ SUBROUTINE 'TRANSITION NOTE -> FREQUENCY' +++
00480 :
00490 : IF R$ = "H0" THEN LET R = 251 : RETURN
00500 : IF R$ = "C"  THEN LET R = 237 : RETURN
00510 : IF R$ = "C#" THEN LET R = 224 : RETURN
00520 : IF R$ = "D"  THEN LET R = 211 : RETURN
00530 : IF R$ = "D#" THEN LET R = 199 : RETURN
00540 : IF R$ = "E"  THEN LET R = 188 : RETURN
00550 : IF R$ = "F"  THEN LET R = 177 : RETURN
00560 : IF R$ = "F#" THEN LET R = 167 : RETURN
00570 : IF R$ = "G"  THEN LET R = 157 : RETURN
00580 : IF R$ = "G#" THEN LET R = 149 : RETURN
00590 : IF R$ = "A"  THEN LET R = 140 : RETURN
00600 : IF R$ = "B"  THEN LET R = 132 : RETURN
00610 : IF R$ = "H"  THEN LET R = 124 : RETURN
00620 : IF R$ = "C1" THEN R = 117 : RETURN
00630 : IF R$ = "C1#" THEN R = 111 : RETURN
00640 : IF R$ = "D1" THEN R = 104 : RETURN
00650 : IF R$ = "D1#" THEN R = 99 : RETURN
00660 : IF R$ = "E1" THEN R = 93 : RETURN
00670 : IF R$ = "F1" THEN R = 88 : RETURN
00680 : IF R$ = "F1#" THEN R = 83 : RETURN
00690 : IF R$ = "G1" THEN R = 78 : RETURN
00700 : IF R$ = "G1#" THEN R = 73 : RETURN
00710 : IF R$ = "A1" THEN R = 69 : RETURN
00720 : IF R$ = "B1" THEN R = 65 : RETURN
00730 : IF R$ = "H1" THEN R = 61 : RETURN
00740 : IF R$ = "C2" THEN R = 57 : RETURN
00750 : IF R$ = "Z"  THEN R = 0 : RETURN
```

```
00760 :
00770 : RETURN
00780 :
00790 REM === MUSICAL COMPOSITION ==========================
00800 :
00810 DATA A1,2,Z,2,C2,2,A1,2,E1,2,Z,2,A1,2
00820 DATA E1,2,C1,2,Z,2,E1,2,C1,2,A,60
00830 :
00840 : REM INSERT THE REST OF THE NOTES HERE
00850 :
00860 DATA @,0
```

****

The high point of this chapter is the famous Game of Life published by Martin Gardner in his column "Mathematical Games" in the Scientific American (October 1970 and September 1971). Gardner introduced the world to this game invented by John Horton Conway called "Life". The cosmic name derives from the fact the game has similarities with the rise, fall and alteration of populations of living organisms.

Example 3.9   THE GAME OF LIFE
The playing area, a plane divided into quadrants, represents the area where life takes place. Each field can have two conditions, either 0 (unoccupied, or dead) or 1 (occupied, or alive). The neighbors of field j are all the fields within the quadrant with side length 3 and with j as its center. The following transformation rules apply:

```
O   O   O
O   *   O
O   O   O
```

(1) An empty field becomes occupied when exactly three of its neighboring fields are occupied.

(2) An occupied field becomes empty when fewer than two, or more than three, of its neighboring fields are occupied; otherwise it remains occupied.

You may wonder how Conway developed these rules. He wanted to make certain that, on the one hand, not too many entities enjoyed unrestricted growth. But he also did not want too many of them to die after too short a time. In other words, he desired an equitable balance between life and death and wanted to encourage entities that followed one of these genetic laws:

A. Complete extinction after finite period.

B. Stabilization to a permanent (unchangeable) configuration.

C. Periodic alteration

The rules of the game actually produce configurations from all three classes:

A. A single living cell or a combination of two living cells dies after the first move.

B. The following block of four remains stable (unchanged)

C. The following triplet (called a "blinker") alternates with period 1:

But there is even more fascinating behavior and a surprising multiplicity of forms as the following program shows:

```
00100 : PRINT CHR$(147)
00110 : PRINT "        GAME OF LIFE
00120 : PRINT "        ------------
00130 :
00140 REM 'GAME OF LIFE' BY J.H.CONWAY
00150 :
00160 : PRINT "THIS GAME SIMULATES THE RISE, FALL AND CHANGE
00170 : PRINT "OF POPULATIONS OF ANIMATE AND INANIMATE
00175 : PRINT "ORGANISMS.
00180 :
00190 : PRINT : PRINT "PRESS A KEY.
00200 :
00210 : GET T$ : IF T$ = "" THEN 210
00220 :
00230 REM *** MAIN PROGRAM ******************************
00240 :
00250 : GOSUB 380   : REM INITIALIZATION
00260 : GOSUB 510   : REM ENTER
00270 : GOSUB 740   : REM DETERMINING THE NEXT GENERATION
00280 : GOSUB 1170  : REM DISPLAY
00290 :
00300 : PRINT "ANOTHER GENERATION?
00310 : PRINT "THEN HIT SPACE BAR.
00320 : GET T$ : IF T$ = "" THEN 320
00330 : IF T$ = " " THEN 270
```

```
00340 : END
00350 :
00360 REM *** END OF MAIN PROGRAM **************************
00370 :
00380 REM +++ SUBROUTINE 'INITIALIZATION' +++++++++++++++++++
00390 :
00400 : LET N = 20          : REM SIDE LENGTH OF LIVING SPACE
00410 :
00420 : DIM A(N+1,N+1), B(N+1,N+1) : REM TWO COPIES OF THE
00425 :                             REM LIVING SPACE
00430 :
00440 : LET CR$ = CHR$(29) : LET CD$ = CHR$(17)
00450 : FOR I = 1 TO 6
00460 :    LET CR$ = CR$ + CR$ : LET CD$ = CD$ + CD$
00470 : NEXT I
00480 :
00490 : RETURN
00500 :
00510 REM +++ SUBROUTINE 'ENTER' ++++++++++++++++++++++++++++
00520 :
00530 : GOSUB 1170 : REM DISPLAY
00540 :
00550 : LET M = N*N
00560 :
00570 : PRINT CHR$(19) LEFT$(CD$,22);
00580 : PRINT "ENTER ROW AND COLUMN, END WITH '--'";
00590 :
00600 : FOR K = 1 TO M
00610 :    PRINT CHR$(19) LEFT$(CD$,23);
00620 :    FOR I = 1 TO 39 : PRINT " "; : NEXT I
00625 :    PRINT CHR$(13) CHR$(145);
00630 :    INPUT "ROW: "; I$
00640 :    IF I$ = "--" THEN RETURN
00650 :    LET I = VAL(I$)
00660 :    PRINT CHR$(145) TAB(15);
00670 :    INPUT "COLUMN: "; J
00680 :    LET A(I,J) = 1
00690 :    PRINT CHR$(19) LEFT$(CD$,I) LEFT$(CR$,J-1); "O"
00700 : NEXT K
00710 :
00720 : RETURN
00730 :
00740 REM SUBROUTINE 'DETERMINING THE NEXT GENERATION'
00750 :
00760 :    FOR I = 1 TO N
00770 :      FOR J = 1 TO N
00780 :        GOSUB 1050 : REM DETERMINING NUMBER OF
00785 :                     REM NEIGHBORS
00790 :
00800 :        IF A(I,J) = 1 THEN 870 : REM CELL I,J IS
00805 :                                  REM OCCUPIED
00810 :
```

91

```
00820 :     REM --- BIRTH?
00830 :
00840 :        IF NZ = 3 THEN LET B(I,J) = 1 : GOTO 920
00850 :        GOTO 900
00860 :
00870 :     REM --- DEATH?
00880 :
00890 :        IF NZ = 2 OR NZ = 3 THEN B(I,J) = 1 : GOTO 920
00900 :        LET B(I,J) = 0
00910 :
00920 :     NEXT J
00930 :   NEXT I
00940 :
00950 : REM --- NEW GENERATION BECOMES OLD ONE
00960 :
00970 :   FOR I = 1 TO N
00980 :     FOR J = 1 TO N
00990 :       LET A(I,J) = B(I,J)
01000 :     NEXT J
01010 :   NEXT I
01020 :
01030 : RETURN
01040 :
01050 REM +++ SUBROUTINE 'NUMBER OF NEIGHBORS' +++++++++++++
01060 :
01070 : LET NZ = 0
01080 : FOR K = I-1 TO I+1
01090 :    FOR L = J-1 TO J+1
01100 :       LET NZ = NZ + A(K,L)
01110 :    NEXT L
01120 : NEXT K
01130 : LET NZ = NZ - A(I,J)
01140 :
01150 : RETURN
01160 :
01170 REM +++ SUBROUTINE 'DISPLAY' ++++++++++++++++++++++++++
01180 :
01190 : PRINT CHR$(147)
01200 : FOR I = 1 TO N
01210 :    FOR J = 1 TO N
01220 :       IF A(I,J) = 1 THEN PRINT "O"; : GOTO 1240
01230 :                        PRINT ".";
01240 :    NEXT J
01250 :    PRINT
01260 :
01270 : NEXT I
01280 :
01290 : RETURN
01300 :
```

An interesting form in this 'game' is the "glider",
which is a configuration that reproduces itself in the fifth
generation, but also glides one field to the right and one
field down.

```
. . . . .   . . . . .   . . . . .   . . . . .   . . . . .
. . O . .   . . . . .   . . . . .   . . . . .   . . . . .
. . . O .   . O . O .   . . . O .   . . O . .   . . . O .
. O O O .   . . O O .   . O . O .   . . . O O   . . . . O
. . . . .   . . O . .   . . O O .   . . O O .   . . O O O
```

Unfortunately, this program has a small problem. It is un-
bearably slow. It can be speeded up somewhat (see the
brainteasers at the end of the chapter) but, ultimately,
real help comes from programming in machine language. Because of lack of space, we can't go into the area of machine
language here. Nonetheless, this example will give you an
idea of what it is about. If you become inspired to pursue
this topic, let me assure you that you'll find it fascinating.

We communicate with the computer in the programming
language BASIC which is a so-called "higher programming
language." This simply means that is oriented to human
language and is thus "user friendly", as opposed to being
oriented to the binary "guts" of the machine. BASIC's advantage is that it is very comprehensible; it's disadvantage
is that first it must be translated into the machine's language (machine language). In reality, the computer only
understands bit patterns (series such as 10101001). To
translate from BASIC into this 0-1 series, the computer uses
a voluminous program called an "Interpreter." Translating a
command into this machine language doesn't take much time,
but when there are very many commands to be executed, the
lag time can be noticeable -- as in the Game of Life. A
program written initially in machine language would save
lots of translation time but it does have the disadvantage
of being agonizingly difficult to write. Take a look at the
number series in the DATA lines from 810 on in the following
program.

```
00100 : PRINT CHR$(147)
00110 : PRINT "       LIFE
00120 : PRINT "       ----
00130 :
```

93

```
00140 REM CONWAY'S 'GAME OF LIFE' IN MACHINE LANGUAGE
00150 :
00160 : PRINT : PRINT
00170 : PRINT "THIS GAME SIMULATES THE RISE, FALL AND CHANGE
00180 : PRINT "OF POPULATIONS OF ANIMATE OR INANIMATE
00185 : PRINT "GROUPS.
00190 :
00200 : PRINT : PRINT "        PRESS A KEY.
00210 :
00220 : GET T$ : IF T$ = "" THEN 220
00230 :
00240 REM *** MAIN PROGRAM *******************************
00250 :
00260 : GOSUB 340      : REM INITIALIZATION
00270 :
00280 : GOSUB 460      : REM INPUT
00290 :
00300 : GOSUB 700      : REM CALCULATING THE NEXT GENERATION
00310 :
00320 : GOTO 300       : REM CONTINUOUS LOOP
00330 :
00340 REM +++ SUBROUTINE 'INITIALIZATION' +++++++++++++++++
00350 :
00360 : PRINT
00370 : PRINT "HOW MANY SECONDS SHALL A GENERATION REMAIN
00380 : INPUT "ON THE SCREEN? ";T
00390 :
00400 : REM ENTERING THE MACHINE PROGRAM
00410 :
00420 :     FOR I = 826 TO 950 : READ C : POKE I,C : NEXT I
00430 :
00440 :     RETURN : REM END OF INITIALIZATION
00450 :
00460 REM +++ SUBROUTINE 'INPUT' +++++++++++++++++++++++++++
00470 :
00480 : PRINT
00490 : PRINT "ENTER THE BEGINNING FIGURE BY USING THE
00500 : PRINT "CURSOR KEY FOR THE LIVING CELLS 'O' AND FOR
00510 : PRINT "YOUR POSITION. BEGIN THE REPRODUCTION WITH
```

```
00515 : PRINT "      " CHR$(18) "<RETURN>" CHR$(146)
00520 :
00530 : PRINT : PRINT : PRINT "     PRESS A KEY "
00540 : GET A$ : IF A$ = "" THEN 540
00550 : PRINT CHR$(147);"#";CHR$(157);
00560 :
00570 : GET A$ : IF A$ = "" THEN 570
00580 :
00590 : IF A$ = CHR$(29) OR A$ = CHR$(157) THEN 640
00595 : IF A$ = CHR$(17) OR A$ = CHR$(145) THEN 640
00600 : IF A$ = "○" OR A$ = CHR$(20) THEN 640
00610 :
00620 : IF A$ = CHR$(13) THEN PRINT "  " : RETURN
00625 :    REM RETURN KEY
00630 :
00640 : PRINT " ";CHR$(157);A$;"#";CHR$(157);
00645 :    REM MARK TO SET FIGURE LOCATION
00650 :
00660 : GOTO 570
00670 :
00680 : REM +++ END OF INPUT +++
00690 :
00700 REM ++ SUBROUTINE 'CALCULATING THE NEXT GENERATION'++
00710 :
00720 : SYS 887 : REM CALL UP MACHINE PROGRAM
00730 :
00740 : LET T2=TI
00750 : IF TI < T2 + 60*T THEN 750 : REM TIME BETWEEN TWO
00755 :                                REM GENERATIONS
00760 :
00770 : RETURN : REM END CALCULATION OF NEXT GENERATION
00780 :
00790 REM === MACHINE PROGRAM ===============================
00800 :
00810 DATA 169,215,133,10,169,127,133
00820 DATA 11,169,215,133,12,169,26,133
00830 DATA 13,96,230,10,208,2,230,11
00840 DATA 230,12,208,2,230,13,165,11
00850 DATA 201,132,208,4,165,10,201,16
00860 DATA 96,169,0,133,14,162,7
00870 DATA 188,175,3,177,10,201
00880 DATA 81,208,2,230,14,202,16,242,96
00890 DATA 32,58,3,32,98,3,160,41
00900 DATA 177,10,201,81,240,10,165
00910 DATA 14,201,3,240,14,169,32,208
00920 DATA 12,165,14,201,3,240,4,201
00930 DATA 2,208,242,169,81,145,12,32
00940 DATA 75,3,208,216,32,58,3,177
00950 DATA 12,145,10,32,75,3,208,247
00960 DATA 96,0,1,2,40,42,80,81,82
00970 :
00980 REM === END OF MACHINE PROGRAM =======================
```

If we wanted to formulate and execute part of the the program in machine language, we would proceed as follows (with CBM-computers).

Step 1: Developing a partial program in assembly language. The beginning of the assembly program for "Life" -- actually, its main program -- goes like this (see below).

887 : 32	JSR	AB	826	: to subroutine "Initialize"
890 : 32	JSR	AB	866	: to subroutine "Count neighbors"
893 : 160	LDY	#	41	: is the field occupied?
895 : 177	LDA	INY	10	:
897 : 201	CMP	#	81	:
899 : 240	BEQ	R	10	: skip, if line is occupied
901 : 165	LDA	ZP	14	: number of neighbors
903 : 201	CMP	#	3	: three neighbors?
905 : 240	BEQ	R	14	: skip, if three neighbors
907 : 169	LDA	#	32	: empty the cell
909 : 208	BNE	R	12	: skip, if not empty
911 : 165	LDA	ZP	14	: number of neighbors
913 : 201	CMP	#	3	: three neighbors?
915 : 240	BEQ	R	4	: if yes, then birth
917 : 201	CMP	#	2	: two neighbors?
919 : 208	BNE	R	42	: if not, then jump back
921 : 169	LDA	#	81	: creates new individual
923 : 145	STA	INY	12	: set in new field
925 : 32	JSR	AB	843	: increment addresses
928 : 208	BNE	R	216	: jump back if field not finished
930 : 32	JSR	AB	826	: to subroutine "Initialize"
933 : 177	LDA	INY	12	: new generation on screen
935 : 145	STA	INY	10	:
937 : 32	JSR	AB	843	: increment field addresses
940 : 208	BNE	R	247	: jump back
942 : 96	RTS			: end of main program

Step 2: Changing the assembly program into a series of machine commands. This series (written in decimal form) for the Life program goes like this: 32, 58, 3, 32, 98, 160, 41, ...

Step 3: Entering the machine program as DATA lines in the BASIC program (in our example beginning with line 810). The main program is stored starting in DATA line 890). Entering data into memory occurs in line 420 of the BASIC program. Every machine command is entered into the variable C and then, using POKE I,C, into memory cell No. I. The machine program is now in cells 826 through 950.

Step 4: Calling up the machine program from the BASIC program.

This occurs in our example using the command SYS 887 (line 720); the main machine program begins in cell 887 (see the list of commands above).

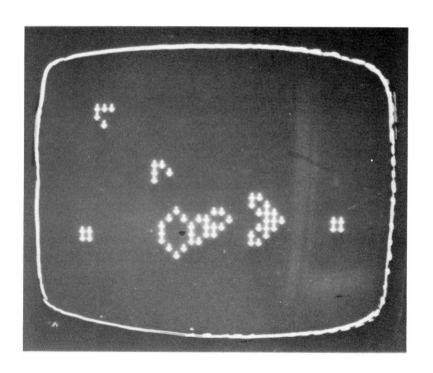

ADDITIONAL BRAINTEASERS

1. Draw quadrangles on your screen after first placing the dots to mark the four corners.

2. Draw a diagonal line between two given points.

3. Write a short program that will draw a trangle after marking the corners with dots.

4. Make a ball (moving point) roll around a rectangle so that it bounces off the side as on a pool table.

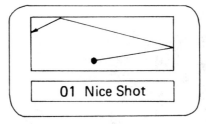

5. Create an "information line" to be displayed on the screen while playing games that will tell the player how many tries he's had and how many times he has scored. Let this line also display the time elapsed.

6. Construct an object out of several characters and then move it around the screen.

7. Display a blinking cursor after the GET command.

8. Write a program for a quick game of "Cat and Mouse": randomly distributed points will appear on the screen. These are then "eaten" by a character you control. Give your cat a lot of speed.

9. Build a time factor into the program "Sonar Sensing" (The Bat) that will give a player a survival time. He has to

play against the clock and wins if he can beat the survival time.

10. Make the survival time in "Sonar Sensing" (The Bat) variable so you can select different degrees of difficulty.

11. The degree of difficulty in "Sonar Sensing" can be changed another way. Try altering the time factor (LET N = 0.98 * N in line 890).

12. You can award points that are proportional to how well a player does in relation to the time elapsed in the game. The player who finishes before the time is up gets a higher score; a slower time earns fewer points. Build this feature into the program for "Sonar Sensing."

13. Display on the screen a line showing the time elapsed as well as the time left in the game.

14. Let the computer play in "Sonar Sensing" as an opponent. Program a random path for the computer so it can avoid obstacles. But don't make the computer too perfect or the human players will give up.

15. Design a game analagous to "Sonar Sensing" in which obstructions must be avoided; but in this version your dot has to end up at a pre-set target spot.

16. Program a game in which a "snake" moves across the screen "eating" objects and getting longer with each bite. The player whose snake is longest is the winner. But 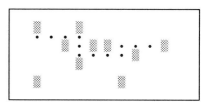 if a player's snake eats a "poisoned" object, it gives up the ghost and the game is lost.

17. In the game "Star Hunt" program the shots to leave a trail behind them: —▸ – – – – – – – – – – – *

18. Make the obstacles in program "Sonar Sensing" multiply more and more quickly so the point has to move faster to avoid them.

19. Put stars of varying point value into the game "Star Hunt." Exploding the larger ones earns a higher score.

20. Program "Star Hunt" like "Sonar Sensing" so that a player can select various degrees of difficulty.

21. Change the game "The Lizard's Snack" so that the fly does not remain stationary, but moves at random and makes different noises. Add sound effects to the lizard's activity. Try different noises for catching and for eating the flies.

22. Here is a mini-game: a player shoots at targets using a rotatable "cannon."

Sorry for all the shooting. From now on things stay pretty peaceful.

23. You can improve the mouse's strategy in "The Maze" if the computer learns to avoid the dead ends. Of course, with this ability no mortal player can beat the machine but try to program it anyway.

24. Write a search program to escape from Sam Lloyd's number maze (see 3.7). How many paths are there out of the woods? Sam Lloyd said there was only one, but was he right? Sam's solution: SW to 4, SW to 6, NE to 6, NE to 2, NE to 5, SW to 4, SW to 4, SW to 4, and then the final step to NW and freedom.

25. Another solution for getting out of the number maze goes like this: N to 2, E to 4, S to 1, E to 2, S to 2, S to 2, W

to 2, W to 2, NW to 4. Now we are are at the point (line 14
in the diagram, the 7th element from the left) where Lloyd's
path also leads. All the solutions that differ from Lloyd's
proceed across the 2 in the 6th line from the bottom, 8th
element from the left. Demonstrate that if you replace this
2 with another number not equal to 0, then only Lloyd's
solution is possible. (Maybe the artist who drew the graph
made a mistake and Sam Lloyd was right.)

26. Program the computer to make random steps. What is the
probability that it will get out of the maze?

27. Transpose other
pieces of music into
computer music as you
learned in example 3.8.

28. Expand the "Number
Sensor" so that every
number that appears on
the screen sounds
a different tone.

29. Program a game like
"Total Recall" that asks
a player to reproduce short tunes from memory.

30. Speed up the "Game of Life" by pinpointing the relevant
area in which changes (birth or death) are to be expected
before determining the neighbors.

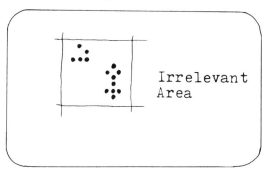

31. In the "Game of Life" examine the development of the following figures:

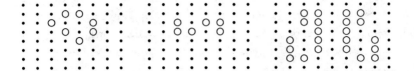

32. The eater devours a glider:

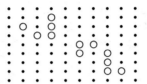

Try this one out!

33. In this one a glider also gets eaten:

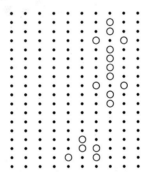

Let this process run on the screen.

34. This is the famous "Glider Cannon" that has led to speculation in the field of computer science.

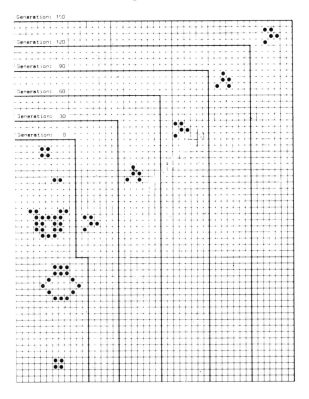

35. Experiment with the "Game of Life." Try it with other rules of reproduction. For example, let birth be possible when there are two or three living neighbors.

36. Another game of reproduction was developed by the mathematician Stanislav Ulam. He defines a so-called orthogonal proximity as depicted here. A cell has only four neighbors arranged like a cross. A cell is born when exactly one living neighbor exists. Every cell dies after its second generation. This game produces incredible figures that keep growing and changing. Write a program for it.

CHAPTER IV

GAMES OF SEARCHING AND GUESSING

Every one of us spends a great portion of our lives looking for things. Just think of looking for articles of clothing or for utensils in a department store, for good programs on television and for all those misplaced or lost items. Or consider the search for just the right word when speaking or an efficient algorithm when calculating. To put it more generally, every act of problem solving involves searching. In a sense all goal-directed activity can be categorized as "Searching."

Typical problem situations are presented in games in simplified form. The object of play is to execute a search efficiently, that is, with as few tests as possible. Tests are information-gathering searches about the solution one is looking for.

The computer can play a two-fold role in this process. It can pose the problem and evaluate the guesses. Or, it can do the guessing itself -- as long as it is given the strategy for guessing and searching. In our first example the computer will be asked to guess a number thought of by the player. The remarkable thing about this game is that the computer seems to get by without knowing enough information. However, it can be shown mathematically that the information is adequate.

Example 4.1 MATHLETE (Nichomachus)
The player thinks of a number between 1 and 1000. The computer then asks him to divide his number by 7, 11, and 13 and to report how much is left over. The computer then figures out the number thought of.

Dialogue:

```
How much is left over after dividing by 7? 2
How much is left over after dividing by 11? 3
How much is left over after dividing by 13? 6
Let me think about this one. . . .
Were you thinking of the number 58? yes
Would you like to know how I figured that out?
Then. . .
```

It is true that 58 = 8*7+2 = 5*11+3 = 4*13+6. But how did
the computer arrive at 58? It did it just by calculating

$$z = 715*2 + 364*3 + 924*6 = 8066$$

and then got how much was left over after division by 1001
= 7*11*13. The remainder is the number the player was
thinking of. The theory behind all this is prefigured in
the work of the Greek mathematician Nichomachus of Gerasa
(c. 100 B.C.).

```
00100 : PRINT CHR$(147)
00110 : PRINT "        MATHLETE
00120 : PRINT "        --------
00130 :
00140 REM GUESSING NUMBERS USING THE CHINESE
00145 REM REMAINDER THEOREM
00150 :
00160 : PRINT
00170 : PRINT "THINK OF A NUMBER BETWEEN 1 AND 1000.
00180 : PRINT "I SHALL GUESS IT ONCE YOU HAVE ANSWERED
00190 : PRINT "3 LITTLE QUESTIONS.  YOU WILL HAVE TO
00200 : PRINT "STATE THE REMAINDER WHEN YOUR NUMBER IS
00210 : PRINT "DIVIDED BY 7, 11, AND 13.
00220 :
00230 : PRINT
00240 : INPUT "REMAINDER AFTER DIVISION BY  7 "; R1
00250 :    IF R1 < 0 OR R1 > 6 THEN 240
00260 : INPUT "REMAINDER AFTER DIVISION BY 11 "; R2
```

```
00270 :    IF R2 < 0 OR R2 > 10 THEN 260
00280 : INPUT "REMAINDER AFTER DIVISION BY 13 "; R3
00290 :    IF R3 < 0 OR R3 > 12 THEN 280
00300 :
00310 : PRINT
00320 : PRINT "I SHALL HAVE TO THINK A MOMENT. . .
00330 : FOR I = 1 TO 2000 : NEXT I
00340 :
00350 : LET Z = 715*R1 + 364*R2 + 924*R3
00360 : IF Z > 1001 THEN LET Z = Z-1001 : GOTO 360
00370 :
00380 : PRINT
00390 : PRINT "WERE YOU THINKING OF THE NUMBER ";Z;"?(Y/N)
00400 :
00410 : GET A$ : IF A$ = "" THEN 410
00420 : IF A$ = "Y" THEN 460
00430 : IF A$ = "N" THEN 520
00440 : PRINT "I AM AFRAID I DID NOT UNDERSTAND." : GOTO 410
00450 :
00460 : PRINT
00470 : PRINT "CARE TO KNOW HOW I FIGURED IT OUT?
00480 : PRINT "THEN LOOK IT UP IN THE INTRODUCTION TO
00490 : PRINT "ARITHMETIC BY THE ANCIENT GREEK NICHOMACHUS.
00500 : END
00510 :
00520 : PRINT
00530 : PRINT "THERE IS SOMETHING WRONG WITH YOUR
00540 : PRINT "ARITHMETIC.  I SUGGEST ANOTHER TRY.
00550 :
00560 : FOR I = 1 TO 2000 : NEXT I
00570 : RUN
```

After that little excursion into number theory let's turn to guessing games in the true sense of the word.

Example 4.2 GUESS THE NUMBER
The object is to find a randomly chosen integer between
1 and M by repeatedly defin-
ing the region in which you
think it lies.

Dialogue:

```
I am thinking of a number between 1 and 352.
Give me the boundaries of the region:
      Lower boundary? 1
      Upper boundary? 170
The number is within that region.
      Lower boundary? 85
      Upper boundary? 170
The number is not within that region.
      Lower boundary?
  (etc................)
      Lower boundary? 69
      Upper boundary? 69
YOU GUESSED IT
The number was 69. Good-bye.
```

This program is probably pretty easy to understand by now:

```
00100 : PRINT CHR$(147)
00110 : PRINT "         GUESS THE NUMBER
00120 : PRINT "         ---------------
00130 :
00140 REM NUMBER GUESSING WHEN SEARCH REGIONS ARE DEFINED
00150 :
00160 : LET M = INT(10000*RND(TI)+1)
00170 :
00180 : PRINT
00190 : PRINT "I AM THINKING OF A NUMBER ";
00195 : PRINT "BETWEEN 1 AND ";M;".
00200 : PRINT
00210 : PRINT "YOUR OBJECT IS TO FIND THE NUMBER AFTER
00220 : PRINT "REPEATEDLY DEFINING A REGION, I.E., AN UPPER
00230 : PRINT "AND LOWER BOUNDARY FOR YOUR NUMBER.
00240 :
00250 : LET X = INT(M*RND(TI)+1)
00260 :
00270 : PRINT
00280 : PRINT "ENTER THE BOUNDARY NUMBERS OF THE REGION
00290 : PRINT
00300 : INPUT "LOWER BOUNDARY "; L
00310 : INPUT "UPPER BOUNDARY "; U
00320 : IF U < L THEN PRINT "NOT PERMITTED." : GOTO 290
00330 : IF U = L THEN 410 : REM REGION BOUNDARIES IDENTICAL
00340 :
00350 : PRINT : PRINT
```

```
00360 : PRINT "THE NUMBER IN QUESTION DOES ";
00370 :
00380 : IF X < L OR X > U THEN PRINT "NOT ";
00390 : PRINT "LIE WITHIN THE REGION." : GOTO 290
00400 :
00410 : PRINT CHR$(147); CHR$(17); CHR$(147); TAB(10);
00420 : IF L <> X THEN PRINT "THAT'S WRONG!!" : GOTO 290
00430 : PRINT CHR$(18); "YOU GUESSED IT!"; CHR$(146)
00440 : PRINT CHR$(17) "THE NUMBER IN QUESTION IS "; X
00450 :
00460 : END
```

Example 4.3   GUESS THE WORD

You are to guess a word randomly chosen by the computer.
You may guess letters you suspect are in the word. If they
are there, they will be displayed in the right place. You
can also guess whole words.

Dialogue:

```
Game word: . . . . .
Guess a letter or a word: p
Game word: . p . . .
Guess a letter or a word: a
Game word: . p . . a
Guess a letter or a word: opera
Terrific!  You guessed it.
```

The accompanying program is constructed as follows:
First a random number Z is chosen between 1 and 14. This is
the number of words in the DATA statements in the program.
(Of course, you will want to include a lot more words.) The
word corresponding to this number is then entered. After
its length L is determined, the computer displays a "help
line" with the same number of dots as there are letters in
the word. After the player enters the letter he is guessing
(L$), the unknown word U$ is searched letter by letter and
compared with L$. The new clue word N$ is simultaneously
created. If L$ appears in the unknown word, it is inserted
into the clue word. This takes place in the part called
"Constructing the Clue Word" from line 410 on. If a player
guesses an entire word or phrase (line 380) then this is

109

compared with the unknown word (or phrase) and the appropriate message is displayed.

```
00100 : PRINT CHR$(147)
00110 : PRINT "        GUESS THE WORD
00120 : PRINT "        --------------
00130 :
00140 REM GAME FOR LOVERS OF FAIRY TALES
00145 REM AND CHILDREN'S BOOKS
00150 :
00160 REM === RULES OF PLAY =================================
00170 :
00180 : PRINT : PRINT : PRINT "I, YOUR COMPUTER, AM
00190 : PRINT "THINKING OF A WORD YOU MUST GUESS.  YOU MAY
00195 : PRINT "ENTER ONE LETTER AT A TIME. THESE WILL
00200 : PRINT "APPEAR IN THEIR CORRECT SPOT IN THE WORD.
00210 : PRINT "WHEN YOU KNOW THE WORD, ENTER IT ALL.
00220 :
00230 REM === SELECTING THE UNKNOWN WORD ===================
00240 :
00250 : LET Z = INT(14*RND(TI))+1 : REM CHOOSE RANDOM NUMBER
00260 : FOR I = 1 TO Z : READ U$ : NEXT I
00265 :    REM READ THROUGH WORD Z
00270 :
00280 REM === CONSTRUCTING FIRST CLUE WORD =================
00290 :
00300 : LET L = LEN(U$)           : REM NUMBER OF LETTERS
00310 : LET H$ = ""
00320 : FOR I = 1 TO L : LET H$ = H$ + "." : NEXT I
00330 :
00340 REM === GUESS ========================================
00350 :
00360 : PRINT "YOU ARE LOOKING FOR: "; H$
00370 : PRINT : INPUT "YOUR GUESS: LETTERS OR WORD "; L$
00380 : IF L$ = U$ THEN 530 : REM WORD GUESSED CORRECTLY
00390 : IF LEN(L$) > 1 THEN 340
00395 :    REM WORD GUESSED INCORRECTLY
00400 :
00410 REM === CONSTRUCTING THE CLUE WORD ===================
00420 :
00430 : LET N$ = ""               : REM NEW CLUE
00440 : FOR I = 1 TO L
00450 :    LET B$ = MID$(U$,I,1) : REM 'I'TH LETTER OF THE
00455 :                            REM UNKNOWN WORD
00460 :    LET C$ = MID$(H$,I,1) : REM LETTER OF CLUE WORD
00470 :    IF L$ = B$ THEN N$ = N$+L$ : GOTO 490
00475 :       REM LETTER ENTERED
00480 :    LET N$ = N$ + C$       : REM CLUE LETTER COPIED
00490 : NEXT I
00500 : LET H$ = N$            : REM NEW CLUE WORD BECOMES OLD
00510 : GOTO 340                 : REM NEXT TRY
00520 :
00530 REM === EXIT OR REPEAT ===============================
```

```
00540 :
00545 : PRINT : PRINT
00550 : PRINT "TERRIFIC! YOU GUESSED IT.
00555 : PRINT : PRINT
00560 : PRINT "ANOTHER TRY? (Y/N)
00570 : GET A$ : IF A$ = "" THEN 570
00580 : IF A$ = "Y" THEN RUN
00590 :
00600 REM === ITEMS TO BE GUESSED =========================
00610 :
00620 DATA RAPUNZEL, CINDERELLA, RUMPELSTILZKIN,
00630 DATA SNOW WHITE, THE FROG PRINCE, SLEEPING BEAUTY
00640 DATA JACK THE GIANT KILLER, RAGGEDY ANN, BOBBSY TWINS
00650 DATA HARDY BOYS, CURIOUS GEORGE, HEIDI, HANS BRINKER
00660 DATA PETER PAN, TOM THUMB, WINNIE THE POOH
00670 :
00680 END
```

Example 4.4   SCRAMBLED WORDS
A word, the letters of which have been scrambled, is displayed quickly on the screen.   The player has to guess it
but also gets a hint.

Dialogue:

```
Hint: The blue planet
    T H R E A
Your answer? earth
Right!
```

Here is the process:
First of all, one of the words is chosen that has been stored at the end of the program in the DATA lines.   Then a
random permutation is created $p(1)$, $p(2)$, . . . $p(L)$ of the
numbers 1, 2,..., L. L is the length (number of letters) of
the word a player must guess.   It goes like this: for all
$i=1,2...L$, $p(i)$ is a random number between 1 and L.   If such
a number has already been assigned, then a new one is created:

```
450 FOR J=1 TO I-1
460    IF (P(I)=P(I-J)) THEN 440
470 NEXT J
```

111

Then

$$\boxed{\text{MID\$(W\$,P(I),1)}}$$

is the randomly chosen letter of the word you are guessing
(W\$), and it appears in the ith position on the screen.

It could happen by pure chance that the scrambled word
is identical to the word the player has to guess, but the
probability of this happening is very low (namely 1 over
n!, where n! = 1*2*3*...*n). If we are talking about an
eight letter word (n=8) the probability is 0.0000248, as
long as these letters are all distinct (no letter is repeated).

```
00100 : PRINT CHR$(147)
00110 : PRINT "        SCRAMBLED WORDS
00120 : PRINT "        ----------------
00130 :
00140 REM WORD-GUESSING GAME.  LETTERS OF WORD ARE SCRAMBLED
00150 :
00160 REM === START AND INITIALIZATION =====================
00170 :
00180 : REM PUT THE DIRECTIONS AND DESCRIPTION HERE.
00190 :
00250 : LET N = 25 : REM GREATEST POSSIBLE LENGTH OF A WORD
00260 : DIM P(N)        : REM TABLE OF PERMUTATIONS
00270 :
00280 REM === GENERATING THE WORD TO BE GUESSED ============
00290 :
00300 : REM --- SETTING THE WORD TO BE GUESSED
00310 :
00320 :    LET Z = INT(4*RND(TI)+1)
00330 :    FOR I = 1 TO Z : READ D$,D$ : NEXT I
00335 :      REM READ OVER
00340 :    READ W$,ST$  : REM ENTER
00350 :
00360 :    PRINT : PRINT "CLUE: "; ST$
00370 :
00380 :    FOR K = 1 TO 500 : NEXT K
00385 :      REM DISPLAY TIME FOR CLUE
00390 :
00400 : REM --- SCRAMBLE (RANDOM PERMUTATION)
00410 :
00420 :    LET L = LEN(W$)
00430 :    FOR I = 1 TO L
00440 :      LET P(I) = INT(L*RND(TI)+1)
00450 :      FOR J = 1 TO I-1
00460 :        IF P(I) = P(I-J) THEN 440
00465 :          REM NUMBER ALREADY GENERATED
```

```
00470 :     NEXT J
00480 :     NEXT I
00490 :
00500 : REM --- DISPLAY SCRAMBLED LETTERS
00510 :
00520 :     PRINT : PRINT : PRINT : PRINT
00530 :     FOR I = 1 TO L : PRINT MID$(W$,P(I),1)" ";
00535 :     NEXT I
00540 :
00550 :     FOR K = 1 TO 1000 : NEXT K : REM PAUSE TO THINK
00560 :
00570 REM === GUESS ========================================
00580 :
00585 : PRINT CHR$(147) : PRINT : PRINT : PRINT : PRINT
00590 : PRINT "YOUR ANSWER?
00600 : INPUT A$
00610 :
00615 : PRINT : PRINT : PRINT : PRINT
00620 : IF A$ = W$ THEN PRINT "CORRECT!" : GOTO 640
00630 :                     PRINT "SORRY, THAT IS WRONG.
00640 : END
00650 :
00660 REM === WORDS AND CLUES =============================
00670 :
00690 DATA ,,BELL, INVENTOR OF TELEPHONE
00700 DATA CLAVICLE, COLLAR BONE
00710 DATA PUCCINI, COMPOSER OF LA BOHEME
00720 DATA PRAGUE, CAPITOL OF CZECHOSLOVAKIA
00730 :
00740 REM ADD YOUR OWN DATA HERE
00750 :
```

****

Example 4.5 CRACK THE CODE

This game is also known as "Super Brain."   The player has
to guess a four digit number even though he is given limited
information.   If a player guesses the right digit in the
correct place, the message "Right!" appears.   If the digit
is right but out of place, then the message "Almost!" appears.   The computer chooses the numbers; the player
guesses.

Dialogue:

```
1st Try: What is your Number?   1234
You got 0 "right" and 1 "almost."
2nd Try:  What is your number?   @
The unknown number was 5619.
```

Here is the process:

First, we generate a random permutation of the numbers from 1 through 9, using a slightly different process from the previous example. Then the (four digit) number entered by the player is examined for correspondence with the number sought. To do this we copy the unknown numbers in table $U(1),...,U(4)$ into a table $C(1),...,C(4)$ and write the player's answers into a table $L(1),..., L(4)$. Then we see how often $L(i) = C(i)$; this is the number of exact correspondances (line 660-700). The number of "almosts" can be ascertained by finding out how often $L(i) = C(j)$ for a pair $i,j \in \{1,2,3,4\}$ (lines 730-800 of the program that follows).

It should be added that a player can always find out the number by simply typing '@' (see the display above). Add this detail to line 220 of the program.

```
00090 : PRINT CHR$(147)
00100 : PRINT CHR$(18) "          CRACK THE CODE
00110 :
00120 REM THE COMPUTER PLAYS THE GUESSING GAME 'SUPERBRAIN'
00130 :
00140 REM === RULES OF PLAY ==================================
00150 :
00160 : PRINT
00170 : PRINT "I, YOUR COMPUTER, AM THINKING OF A 4 DIGIT
00180 : PRINT "NUMBER, EACH DIGIT BEING DIFFERENT. YOU MUST
00190 : PRINT "GUESS THESE. WHEN YOU GUESS A NUMBER IN THE
00200 : PRINT "RIGHT SPOT, THIS IS A 'HIT.'  WHEN THE NUMBER
00210 : PRINT "IS RIGHT BUT THE POSITION IS WRONG, IT COUNTS
00215 : PRINT "AS 'ALMOST.'
00220 :
00230 REM === DETERMINING THE NUMBER TO BE GUESSED =========
00240 :
00250 : FOR I = 1 TO 9 : LET U(I) = I : NEXT I
00260 : FOR I = 9 TO 2 STEP -1
00270 :    LET K = INT(I*RND(TI))
00280 :    LET H = U(I) : LET U(I) = U(K) : LET U(K) = H
00290 : NEXT I
```

```
00300 :
00310 REM === DIALOGUE =====================================
00320 :
00330 : LET N = 0              : REM NUMBER OF GUESSES
00350 : LET N = N+1
00360 :   PRINT : PRINT
00370 :   PRINT N;". GUESS: ";
00380 :   INPUT "WHAT IS YOUR NUMBER"; L$
00390 :   IF L$ = "@" THEN 530 : REM PLAYER GIVES UP
00400 :   LET L$ = LEFT$(L$,4)
00410 :
00420 :   GOSUB 590 : REM DETERMINING A HIT
00430 :
00440 :   IF V = 4 THEN 510   : REM NUMBER GUESSED
00450 :
00460 :   PRINT
00470 :   PRINT " YOU RACKED UP ";V;" HITS ";
00480 :   PRINT "AND ";H;" ALMOSTS."
00490 : GOTO 350
00500 :
00510 : PRINT : PRINT
00520 : PRINT "YOU GUESSED THE UNKNOWN NUMBER IN ";N;
00530 : PRINT "TRIES" : PRINT
00540 : PRINT "THE UNKNOWN NUMBER WAS: ";
00550 : FOR I = 1 TO 4 : PRINT U(I); : NEXT I
00560 :
00570 : END : REM === END OF DIALOGUE =====================
00580 :
00590 REM +++ SUBROUTINE DETERMINING A HIT +++++++++++++++++
00600 :
00610 : FOR I = 1 TO 4
00620 :   LET L(I) = VAL(MID$(L$,I,1))
00630 :   LET C(I) = U(I)
00640 : NEXT I
00650 :
00660 : LET V = 0  : REM NUMBER OF HITS
00670 : FOR I = 1 TO 4
00680 :   IF L(I) <> C(I) THEN 710
00690 :     LET V = V+1
00700 :     LET L(I) = -1 : LET C(I) = -2
00710 : NEXT I
00720 :
00730 : LET H = 0  : REM NUMBER OF ALMOSTS
00740 : FOR I = 1 TO 4
00750 :   FOR J = 1 TO 4
00760 :     IF L(I) <> C(J) THEN 790
00770 :       LET L(I) = -1 : LET C(J) = -2
00780 :       LET H = H+1
00790 :   NEXT J
00800 : NEXT I
00810 :
00820 : RETURN
```

115

Example 4.6 SPY RING
Four spies are hiding in a grid
measuring 10x10. Your object is
to track them down. The computer
will betray the location of a spy
whose cover is blown and will tell
you how far away you are from the
ones still under cover.

Dialogue:

> At the moment the spies are well hidden.
> 1. Guess where a spy is hiding.
> x=? 5
> y=? 5
> The distance to spy no. 1 is 1.4 lengths.
> The distance to spy no. 2 is 2 lengths.
> The distance to spy no. 3 is 5 lengths.
> The distance to spy no. 4 is 1 lengths.

From this we can gather the following information about the
spies' hiding places:

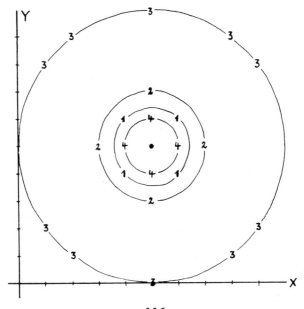

The dialogue continues like this:

```
2. Guess where a spy is hiding.
x=? 5
y=? 4
The distance to spy no. 1 is 1 lengths.
The distance to spy no. 2 is 3 lengths.
The distance to spy no. 3 is 5.8 lengths.
You have captured spy no. 4!
3. Guess where a spy is hiding.
x=?
```

As usual, this program is divided into sections: (1) Directions, (2) Setting the Beginning Values, (here: hiding the
spies), (3) Game and (4) Repetition or Exit. Part three
looks like this:

```
FOR I = 1 TO 10 REPEAT
    Subroutine "Guess"
    IF all spies are found THEN
            Z$:= "Round won"
            Leave subroutine "game"
    END IF
END REPEAT
Subroutine "Round lost"
```

Here is the complete BASIC program for the game:

```
00100 : PRINT CHR$(147)
00110 : PRINT "        SPY RING
00120 : PRINT "        --------
00130 :
00140 REM EASY SEARCHING GAME IN 2 DIMENSIONS
00150 :
00160 REM *** MAIN PROGRAM *******************************
00170 :
00180 : GOSUB 270 : REM GAME RULES
00190 : GOSUB 450 : REM HIDING THE SPIES
00200 : GOSUB 540 : REM ROUND
00210 : GOSUB 1210: REM EXIT OR REPEAT
00220 : END
00230 :
00240 REM *** END OF MAIN PROGRAM ************************
00250 :
```

```
00260 :
00270 REM ### SUBROUTINE 'GAME RULES' ####################
00280 :
00285 : PRINT : PRINT
00290 : PRINT "WOULD YOU LIKE TO SEE THE GAME RULES? (Y/N)
00300 : GET A$ : IF A$ = "" THEN 300
00310 : IF A$ <> "Y" THEN RETURN
00320 :
00325 : PRINT : PRINT
00330 : PRINT "FOUR SPIES ARE HIDDEN IN A GRID MEASURING
00340 : PRINT "10 X 10. DISCOVER THEIR POSITIONS BY MARKING
00350 : PRINT "EACH HIDEOUT WITH A PAIR OF NUMBERS BETWEEN 0
00360 : PRINT "AND 9. I'LL TELL YOU WHEN YOU WERE SUCCESS-
00370 : PRINT "FUL; OTHERWISE, I'LL TELL YOU THE DISTANCE TO
00375 : PRINT "THE SPY." : PRINT : PRINT
00380 : PRINT "YOU HAVE 10 TRIES IN ALL." : PRINT : PRINT
00390 : PRINT CHR$(18) "    PRESS ANY KEY.
00400 :
00410 : GET T$ : IF T$ = "" THEN 410
00420 :
00430 : RETURN
00440 :
00450 REM ### SUBROUTINE 'HIDING THE SPIES' ##############
00460 :
00470 : FOR I = 1 TO 4
00480 :    LET X(I) = INT(10*RND(TI))
00490 :    LET Y(I) = INT(10*RND(TI))
00500 : NEXT I
00510 :
00520 : RETURN
00530 :
00540 REM ### SUBROUTINE 'ROUND' ########################
00550 :
00560 : FOR J = 1 TO 10
00570 :    GOSUB 680                      : REM GUESS
00580 :    GOSUB 920                      : REM ROUND WON?
00590 :    IF Z$ = "WON" THEN RETURN      : REM ROUND OVER
00600 :    PRINT CHR$(18) "PRESS KEY."    : REM PAUSE TO THINK
00610 :    GET T$ : IF T$ = "" THEN 610   : REM CONTINUE
00620 : NEXT J
00630 :
00640 : GOSUB 1040
00650 :
00660 : RETURN : REM END OF 'ROUND' ######################
00670 :
00680 : REM +++ SUBROUTINE 'GUESS' +++++++++++++++++++++++++
00690 :
00695 :    PRINT CHR$(147) : PRINT : PRINT
00700 :    PRINT "THE SPIES ARE NOW WELL HIDDEN. " : PRINT
00710 :    PRINT : PRINT J;". GUESS: FIND A SPY? "
00720 :
00730 :    INPUT " X = "; X
```

118

```
00740 :    IF X < 0 OR X > 9 THEN 730
00750 :    INPUT " Y = "; Y
00760 :    IF Y < 0 OR Y > 9 THEN 750
00770 :
00780 :    FOR I = 1 TO 4
00790 :      IF X(I) = -1 THEN 880 : REM THERE WAS ONE HERE
00795 :                              REM EARLIER
00800 :      IF X(I) <> X OR Y(I) <> Y THEN 850 : REM MISSED
00810 :      PRINT "YOU CAUGHT SPY NO. ";I;"!
00820 :      LET X(I) = -1          : REM MARK POSITION
00830 :      GOTO 880               : REM NEXT SPY
00840 :
00850 :      LET D = SQR((X(I)-X)↑2 + (Y(I)-Y)↑2))
00855 :        REM DISTANCE
00860 :      PRINT "THE DISTANCE TO SPY NO. ";I;" IS";
00870 :      PRINT INT(10*D + 0.5)/10; LENGTHS
00880 :    NEXT I
00890 :
00900 :    RETURN : REM END 'GUESSING' ++++++++++++++++++++
00910 :
00920 : REM +++ 'ROUND WON?' +++++++++++++++++++++++++++++
00930 :
00940 :    FOR I = 1 TO 4
00950 :      IF X(I) = -1 THEN 970
00960 :      RETURN : REM YOU HAVE NOT CAUGHT THEM ALL.
00965 :               REM NOW GUESS.
00970 :    NEXT I
00980 :
00985 : PRINT : PRINT
00990 :    PRINT "YOU HAVE FOUND ALL THE SPIES IN ";J;" TRIES
01000 :    LET Z$ = "WON"
01010 :
01020 :    RETURN : REM END 'ROUND WON?' +++++++++++++++++++
01030 :
01040 : REM +++ SUBROUTINE 'ROUND LOST' +++++++++++++++++++
01050 :
01060 :    PRINT CHR$(147) : PRINT : PRINT
01070 :    PRINT "AFTER 10 TRIES, YOU DID NOT FIND ALL THE
01075 :    PRINT "SPIES. " : PRINT : PRINT
01080 :
01090 :    PRINT "WANT TO KNOW WHERE THEY WERE HIDING? (Y/N)
01100 :    GET A$ : IF A$ = "" THEN 1100
01110 :    IF A$ <> "Y" THEN RETURN
01120 :
01130 :    PRINT : PRINT
01140 :    FOR I = 1 TO 4
01150 :     IF X(I) = -1 THEN 1170
01160 :     PRINT "SPY NO. ";I;" WAS LOCATED
01165 :     PRINT "AT (";X(I);",";Y(I);")"
01170 :    NEXT I
01180 :
01190 :    RETURN : REM END 'ROUND LOST' ++++++++++++++++++
```

```
01200 :
01210 REM ### SUBROUTINE 'EXIT OR REPEAT' #################
01220 :
01230 : PRINT CHR$(147) CHR$(17)
01235 : PRINT : PRINT : PRINT "ANOTHER ROUND? (Y/N)
01240 : GET A$ : IF A$ = "" THEN 1240
01250 : IF A$ = "Y" THEN RUN
01260 :
01270 : RETURN
```

The last example in this chapter combines aspects of Crack
the Code with Battleship.

Example 4.7   LAP (named after L. A. Pijanowski)
Unbeknownst to the player, an 8x8 grid is divided into four
equally large regions. The regions are the same size (the
same number of fields) but different shapes. The goal is to
define the precise boundaries of these regions. To do this
a player guesses in the following way: he defines squares
measuring 2x2 (4 fields); the opponent responds with how
many of the four fields lie in each of the regions.

Let's assume the area is divided into these four regions:

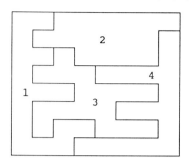

Then the following dialogue takes place:

```
Enter the the test square, but only the coordinates
of its upper left corner.
        Row? 1
        Column? 1
3 field(s) lie in region 1
1 field(s) lie in region 2
0 field(s) lie in region 3
0 field(s) lie in region 4
Do you want to see all the regions? y/n   n
Enter the test square . . .
```

Before we begin with the program, you should be clear about the various possibilities that exist for creating a two-dimensional grid on the computer screen.

1. Coordinates

The method using coordinates has the advantage of traditional mathematical terminology. Each point is designated by a pair of numbers (x,y):

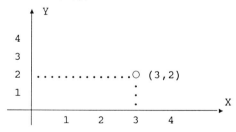

2. Matrix

Using the matrix method one enters row I and column J. The advantage is easy programability. The four fields of the 2x2 square have the following coordinates (row, column):

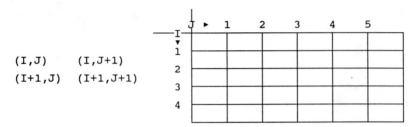

(I,J)   (I,J+1)

(I+1,J)   (I+1,J+1)

3. Chess Board

We'll use the chess board method when we get to the board games. Each point is designated by a pair (letter, number).

In the game LAP we are going to use the matrix method. This way it is easy to enter and store the four regions:

```
FOR I = 1 TO 8
    FOR J = 1 TO 8 : READ F(I,J) : NEXT J
NEXT I
```

In the table M(1),......, M(4) we count how many fields of the test square lie in region K. This goes very simply:

```
FOR R1 = 0 TO 1
    FOR C1 = 0 TO 1
        LET K = F(R+R1,C+C1)
        LET M(K) = M(K) + 1
    NEXT C1
NEXT R1
```

If the player enters the pair R,C as row and column of the upper left corner of his test square, then the loop runs

122

through the fields (R,C), (R,C+1), (R+1,C), (R+1,C+1). The number K = F(R+R1,C+C1) is the number of the region (1, 2, 3, or 4). M(K) counts the number of fields in the given region. If, for example, R,C = 1,1 then K = F(1,1) = 1 and M(1) = M(1) + 1. The following program can be greatly expanded:

```
00100 : PRINT CHR$(147)
00110 : PRINT "        LAP
00120 : PRINT "        ---
00130 :
00140 REM GAME BY L.PIJANOWSKI
00145 REM COPYRIGHT HUGENDUBEL-VERLAG MUNICH
00150 :
00160 REM === START =========================================
00170 :
00180 : REM --- ENTER THE RULES OF PLAY HERE
00190 :
00200 REM === INITIALIZATION ================================
00210 :
00220 : FOR I = 1 TO 8
00230 :    FOR J = 1 TO 8 : READ F(I,J) : NEXT J
00240 : NEXT I
00250 :
00260 DATA 1,1,2,2,2,2,2,2      : REM COMPARE THE DATA
00270 DATA 1,2,2,2,2,2,2,4      : REM LINES WITH THE
00280 DATA 1,1,3,2,2,2,2,4      : REM FOUR REGIONS
00290 DATA 1,3,3,3,4,4,4,4      : REM WITHIN THE
00300 DATA 1,1,1,3,3,3,3,4      : REM SQUARE ABOVE
00310 DATA 1,3,3,3,3,4,4,4
00320 DATA 1,3,1,1,3,3,3,4
00330 DATA 1,1,1,4,4,4,4,4
00340 :
00350 REM === GAME ==========================================
00360 :
00370 : PRINT
00380 : PRINT "ENTER THE TEST SQUARE, BUT ONLY THE COORD";
00390 : PRINT "INATES OF ITS UPPER LEFT CORNER.
00400 : PRINT
00410 : INPUT "ROW         "; R
00420 : INPUT "COLUMN      "; C
00430 :
00440 : PRINT
00450 : PRINT "THESE FIELDS ARE PART OF THE TEST SQUARE ";
00460 : PRINT R;C;"  ";R;C+1;"  ";R+1;C;"  ";R+1;C+1
00470 :
00480 : FOR K = 1 TO 4 : LET M(K) = 0 : NEXT K
00490 :
00500 : FOR R1 = 0 TO 1
00510 :    FOR C1 = 0 TO 1
```

```
00520 :      LET K = F(R+R1,C+C1)
00530 :       LET M(K) = M(K)+1
00540 :    NEXT C1
00550 : NEXT R1
00560 :
00570 : PRINT
00580 : FOR K = 1 TO 4
00590 :   PRINT M(K);" FIELDS ARE IN THE REGION "; K
00600 : NEXT K
00610 :
00620 : PRINT
00630 : PRINT "DO YOU WANT TO SEE ALL THE REGIONS? (Y/N)
00640 : GET A$ : IF A$ = "" THEN 640
00650 : IF A$ <> "Y" THEN 350 : REM NEW ATTEMPT
00660 :
00670 REM === SOLUTION =======================================
00680 :
00690 : PRINT CHR$(147)
00700 : FOR I = 1 TO 8
00710 :   FOR J = 1 TO 8 : PRINT F(I,J);: NEXT J
00720 :   PRINT
00730 : NEXT I
00740 :
00750 : END
```

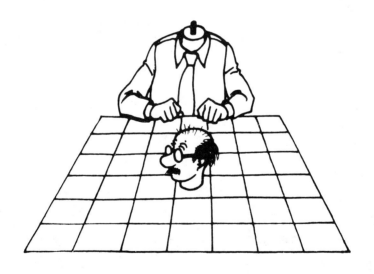

ADDITIONAL BRAINTEASERS

1. In order to figure out the why the mathlete Nichomachus
(example 4.1) seems so miraculous, let's do the following:
a) Enter these remainders: 0,0,1; 0,1,0; 1,0,0.   Then
look at line 350 of the program.
b) Make yourself a table of the remainders when divided by
7, 11, 13:

		Remainder when divided	
Number	by: 7	by 11:	by 13:
500	3	5	6
501	4	6	7
502	5	7	8
503	6	8	9
504	7	9	10
505	8	10	11
506	9	0	12

. . . .

How many such combinations of remainders are there?
c) Try experimenting with numbers other than 7,11,13.

2. Give the player of Guess The Number more information by
telling him whether the number to be guessed is above the
upper boundary or below the lower boundary of the search
region, when the number doesn't fall within the search region.

3. Build a counter into the program for Guess the Number
that will record the tries a player has had and give a
score. You have to consider how many tries are necessary in
the optimal searching strategy to find the unknown number.

Compare this with the number guessing game "Too high -- Too low".

4. Alter the program for 4.3 Guess the Word so that spaces in a hidden phrase are marked with some symbol in the clue word. For example, "Tom Thumb" might appear as:

. . . * . . . . .

5. In the program "Guess the Word", give clues that describe the words or concepts. For example: "What woman let down her hair from a tower?" is a clue to the folktale Rapunzel. Expand the program in this manner.

6. Customize the program "Guess the Word" so that it records the number of guesses and comments on the player's score.

7. HANGMAN
Expand the word guessing game in example 4.3 so that a gallows is gradually erected. With each wrong answer another line of the structure and then a stick figure of the player appear on the screen. A variation on this is to have a picture of a monster gradually take shape that will devour the hapless guesser.

8. Expand the program for Scrambled Words so the player can select both the number of words to guess and their degree of difficulty. Add a score-keeper.

9. Have the scrambled letters in example 4.4 appear at random points around the screen.

10. This is analagous to 4.4. Instead of words appearing on the screen, program proverbs, aphorisms, or famous quotations for the player to decipher. Place the individual words at random points around the screen.

11. Expand the program for Crack the Code so the player can select the length of the number series to be guessed.

12. Let's take a simplified version of Crack the Code in which only the numbers 1,2,3,4 are permitted. Construct an optimal guessing strategy for such a case and program the computer to use it to make the guesses.
Hint: In more than 50% of the cases there is a strategy that makes it possible to figure out the correct number series only after three tries.

13. Expand the program for Spy Ring to include appropriate computer responses to the player who has given incorrect answers. Make them encouraging, such as "Think harder," or "Don't give up when you're so close."

14. TREASURE HUNT
A treasure is hidden on a grid measuring N x N squares. A player is told only a point of the compass (North, South, East, West) to guide his guessing. In a grid measuring 10x10, a player gets only five chances. Now write the program for this game.

15. Expand the program for LAP in such a way that the computer divides the grid into four random regions each consisting of 16 contiguous squares. In addition, have the computer count the tries and keep the score.

CHAPTER V

GAMES OF CHANCE

In games of skill physical prowess or mental faculties
determine success.   In games of chance, however, winning or
losing can depend completely on luck.   Sometimes it is hard
to keep the two types strictly separate.   The games in the
previous chapter were constructed so that the player determined the outcome of the game.   Sometimes winning depended
merely on judging the opportune moment to end the game.
Every computer game of chance needs a "chance generator" to
provide that elusive element of luck.   The oldest chance
generators are perhaps dice, which were favorite playthings
of the Egyptians, Greeks, and Romans.   It is thought that
casting dice was introduced by priests for divination and
prophesying the will of the gods.   Then the practice separa-
ted from religious ritual and became a game all its own.
The Greek historian Herodotus even tells us that Pharaoh
Rhampsinitos descended into the underworld where he played
dice with the goddess Demeter.   "In part he won, in part he
lost; as a gift from Demeter he received a golden cloth."
(Herodutus, Histories, II, 122).

People originally seem to have played dice with anything that would roll and stop in a stationary position.
Shells, sticks, stones and bones were first used for this
purpose.   The earliest dice have 1 and 2, 3 and 4, and 5 and
6 on opposite sides of the cube.   In about 1400 B.C.E. the
modern form came into being. In this "modern" die the dots
of the opposite sides always add up to seven.

Example 5.1 SIX HEX

This is a game, using two dice, for any number of players. The player taking a turn may throw the dice as often as he wants. After stopping, the total number of points thrown accumulates on the player's tally -- as long as no six has turned up. However, if a player throws a six, either on one or both dice (not a total of six) he gets no points for that round and must end his turn. The first player to reach or break 100 wins.

The fun of this game lies in the gamble a player takes when deciding either to keep the points accumulated and relinquish the turn or to keep going for 100 and risk rolling a six.

For the player the question arises after every throw whether to throw again or quit his turn. This decision must be based upon the total points already accumulated by the other players, and from the total accumulated during his current turn. This decision-making process is very complex and hard to explain completely. To get a better idea of this process let's take a closer look at a simplified version of the game. For now we don't care about other opponents, but with the decision, "Shall I throw the two dice again or not?"

Here is the dialogue with the computer, the "chance generator" of the game:

```
Do you want to throw again? (y/n)  y
You have thrown 1 and 4.
Your new subtotal: 5
Do you want to throw again? (y/n)  y
You have thrown 2 and 4.
Your new subtotal: 11
Do you want to throw again? (y/n)  n
Your point total: 11
You could have gotten 23 more points!
Another round? n
Good-bye.
```

How did the computer know how many more points the player could have racked up? The answer is that it plays through the entire series until a six turns up and records the results in two tables R1 (Roll 1) and R2 (Roll 2). The actual game consists simply of accessing this data in the memory and exiting at the right time. In this last case the computer announces the difference S-S1 between the subtotal S1 reached by the player and the total sum S possible during the player's turn. If the player stops too late, that is if the end of the data stored in the memory is reached, then "You lose," appears on the screen (see line 560 of the following program).

```
00100 : PRINT CHR$(147)
00110 : PRINT "        SIX HEX
00120 : PRINT "        -------
00130 :
00140 REM SIMPLIFIED VERSION OF GAME OF CHANCE CALLED PIG
00150 :
00160 REM === RULES OF PLAY ================================
00170 :
00180 : PRINT "ROLL TWO DICE AS OFTEN AS YOU WANT; THE DOTS
00190 : PRINT "ARE ADDED TOGETHER. IF A 6 COMES UP ON ONE OR
00200 : PRINT "BOTH DICE, YOU LOSE THE POINTS YOU HAD";
00205 : PRINT " ROLLED.
00210 :
00220 REM === SERIES OF THROWS ============================
00230 :
00240 : DIM R1(30), R2(30)           : REM TABLES FOR THE ROLLS
00250 :
00260 : LET S = 0                     : REM BEGINNING SCORE
00270 :
00280 : LET I = 1                     : REM ROLL COUNTER
00290 :
00300 : LET R1 = INT(6*RND(TI)+1)     : REM FIRST DIE ROLLED
00310 : LET R2 = INT(6*RND(TI)+1)     : REM SECOND DIE ROLLED
00320 :
00330 : LET R1(I) = R1
00340 : LET R2(I) = R2
00350 :
00360 : IF R1 = 6 OR R2 = 6 THEN 420  : REM SIX HEX TURNS UP
00370 :
00380 : LET S = S + R1 + R2
00390 :
00400 : LET I = I+1 : GOTO 300        : REM NEW ROLL
00410 :
00420 REM === DECISION ON ROLL =========================
00430 :
```

131

```
00440 : LET S1 = 0          : REM TOTAL OF DOTS SHOWING ON DICE
00450 :
00460 : LET I = 1           : REM ROLL COUNTER
00470 :
00480 : PRINT : PRINT "CARE TO ROLL (AGAIN)? (Y/N)
00490 : GET A$ : IF A$ = "" THEN 490
00500 : IF A$ = "N" THEN 630    : REM PLAYER HAS HAD ENOUGH
00510 :
00520 : LET R1 = R1(I)
00530 : LET R2 = R2(I)
00540 :
00550 : PRINT : PRINT "YOU HAVE ROLLED ";R1;" AND ";R2
00560 : IF R1 = 6 OR R2=6 THEN PRINT "YOU LOSE!" : GOTO 680
00570 :
00580 : LET S1 = S1 + R1 + R2
00590 : PRINT "SUBTOTAL ACHIEVED: "; S1
00600 :
00610 : LET I = I+1 : GOTO 480
00620 :
00630 REM === RESULTS =======================================
00640 :
00650 : PRINT : PRINT : PRINT "YOUR POINT TOTAL: "; S1
00660 : PRINT "YOU COULD HAVE GOTTEN "; S-S1;" MORE POINTS.
00670 :
00680 REM === EXIT OR REPEAT ================================
00690 :
00700 : PRINT : PRINT : PRINT "ANOTHER ROUND? (Y/N)
00710 : GET A$ : IF A$ = "" THEN 710
00720 : IF A$ = "Y" THEN RUN
00730 :
00740 : PRINT : PRINT : PRINT "GOOD-BYE.
00750 :
00760 : END
```

After you've played with this program you'll probably discover at what subtotal it seems wise to end a move. This result can be explained theoretically with a simple calculation. Let's assume you have already reached S points and the computer displays "Your new subtotal." If you throw one more time these non-losing numbers could turn up on the dice:

$$2, \quad 3, \quad 4, \quad 5, \quad 6, \quad 7, \quad 8, \quad 9, \quad 10$$

with the following statistical probabilities:

$$\frac{1}{36} \quad \frac{2}{36} \quad \frac{3}{36} \quad \frac{4}{36} \quad \frac{5}{36} \quad \frac{4}{36} \quad \frac{3}{36} \quad \frac{2}{36} \quad \frac{1}{36} \quad .$$

For example, there are six ways to throw a 7 (1,6; 2,5; 3,4;

4,3; 5,2; or 6,1), but two of these ways are losers, so the probability for 7 is 4 (out of 36)!

You chance of losing the accumulated points S is 11/36. The "expectation" is then:

$$\frac{1*2 + 2*3 + 3*4 + 4*5 + 5*6 + 4*7 + 3*8 + 2*9 + 1*10 - 11*S}{36}$$

$$= \frac{150 - 11*S}{36} \ .$$

The value of this term is positive when 150 > 11*S or S < 14. That is, if a player's subtotal is already 14 or more points, he should end his turn. This applies only as long as a player doesn't care how far he is from the 100 point goal or care how many points his opponents already have to their credit.

Now let's consider some points about the final move in the game. Let's assume you have an opponent whose point total is close to 100 and it's your turn. In this case it is better not to apply the rule we have just devised. It is better to roll again in the hope of reaching the goal first. If you already have 98 or 99 points and you roll 2 or more you will have reached or passed that mark. If you roll a 6 on either die you will stay where you already were and it will be your opponent's turn. The probability of winning is 25/36. This is then the way to construct a table showing your statistical chances of winning when working from a certain point level. Things get harder when there are more than two players. You'll find some more directions and suggestions for fun with this game in the Brainteasers at the end of the chapter.

****

Probably the most popular game of dice is craps, or crap shooting. It is a direct descendant of the game "hazard" that was invented by crusaders in the 12th century and very popular in the western world in the 19th century. Players of computer games are always interested to discover that the name of the game probably derives from the English word "crabs", which is an old name for the bad luck rolls of 2, 3, or 12.

Example 5.2 CRAP SHOOT
The game is played with two dice. If the famous "naturals" 7 or 11 appear on the dice, the player automatically wins. But if the old "craps" appear (2, 3, 12) then the player loses. If the player turns up 4,5,6,8,9 or 10 (so-called "points") then the player continues to roll until he hits his point again or rolls a 7. A "point" wins the round; a 7 loses it.

This is played against the bank which rakes in the profits if a player loses and pays out cash if a player wins. A crap game usually takes place on the ground. Picture feverish players throwing their dice against a wall and hunting for the dreaded "snake eyes." In fancy casinos, of course, the game is played on the green felt tables where suave croupiers call the shots and rake in the chips. This is the sort of table used in a gambling casino:

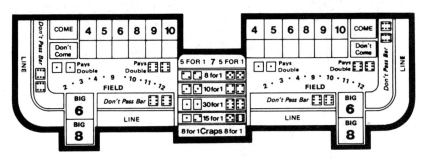

The dice are thrown against the back wall and fall down onto the table where the surface is divided into labelled areas that give betting specifics. "Pass line" means betting on the rolls described above. If you put your money on Big 6 or Big 8 you are betting that these numbers will come up before 7.

The following program is a simulation of betting on Pass line.

```
00100 : PRINT CHR$(147)
00110 : PRINT "        CRAP SHOOT
```

```
00120 : PRINT "          ----------
00130 :
00140 REM GAMBLING WITH DICE
00150 :
00160 REM === BEGINNING CAPITAL AND BET ====================
00170 :
00180 : LET C = 10  : REM PLAYER'S CAPITAL
00190 : LET M = 100 : REM MAXIMUM BET
00200 :
00205 : PRINT : PRINT
00210 : PRINT "YOU HAVE ";C;" DOLLARS IN CASH. ";
00220 :
00230 : INPUT "PLACE YOUR BET? "; E
00240 : IF E <= 0 OR E > C THEN 230
00250 : IF E > M THEN PRINT "BET TOO HIGH." : GOTO 230
00260 :
00270 REM === FIRST ROLL ======================================
00280 :
00290 : LET R = INT(6*RND(TI)) + INT(6*RND(TI)) + 2
00300 :
00310 : FOR J = 1 TO 500 : NEXT J : REM WAITING LOOP
00320 :
00330 : PRINT : PRINT "IN THE FIRST ROLL YOU GOT ";R;
00340 :
00350 : IF R = 7 OR R = 11          THEN 600 : REM WON
00360 : IF R = 2 OR R = 3 OR R = 12 THEN 650 : REM LOST
00370 :
00380 REM === SUBSEQUENT ROLLS ==============================
00390 :
00400 : LET I = 1 : REM ROLL COUNTER
00410 : LET P = R : REM THE 'POINT' TO REMEMBER
00420 :
00430 : PRINT "SECOND SERIES OF ROLLS (YOUR POINT ";
00435 : PRINT "IS";P;"):
00440 :
00450 : LET I = I+1
00460 :
00470 :    LET R = INT(6*RND(TI)) + INT(6*RND(TI)) + 2
00480 :
00490 :    FOR J = 1 TO 300 : NEXT J : REM WAITING LOOP
00500 :
00510 :    PRINT I;". ROLL: ";R;
00520 :
00530 :    IF W = P THEN 570 : REM POINT MADE
00540 :    IF W = 7 THEN 650 : REM LOST
00550 :
00560 : GOTO 450                : REM ANOTHER ROLL
00565 :
00570 :    PRINT "YOU MADE YOUR POINT!"
00575 :
00580 REM === SCORE ==========================================
00590 :
```

```
00600 : PRINT "THAT WINS THE ROUND!
00610 : LET C = C+E               : REM NEW CAPITAL
00620 : PRINT : PRINT "YOU NOW HAVE ";C;" DOLLARS."
00630 : GOTO 690
00640 :
00645 : PRINT : PRINT : PRINT : PRINT
00650 : PRINT "SORRY, YOU LOSE THIS ROUND.
00660 : LET C = C-E           : REM NEW CAPITAL AMOUNT
00670 : IF C <= 0 THEN 750 : REM PLAYER HAS BEEN RUINED
00680 :
00690 REM === GIVE UP OR NEW ROUND ==========================
00700 :
00705 : PRINT : PRINT : PRINT : PRINT
00710 : PRINT "WANT TO TRY ANOTHER ROUND? (Y/N)
00720 : GET A$ : IF A$ = "" THEN 720
00730 : IF A$ = "Y" THEN PRINT CHR$(147) : GOTO 210
00735 :    REM NEW ROUND
00740 :
00750 : PRINT : PRINT : PRINT : PRINT
00755 : PRINT "IT IS BETTER FOR YOU TO QUIT NOW.
00760 :
00770 : END
```

This is how the program runs:

```
You have 10 dollars cash. Place your bet. 2
In the first roll you got 11.
That wins this round!
You have 12 dollars cash. Place your bet. 5
In the first roll you got 4.
Second series of rolls (Your point is 4)
2nd roll: 9
3rd roll: 6
4th roll: 4
You got your point.
That wins the round.
Do you want to try another round? y
You have 17 dollars cash. Place your bet. 17
In the first roll you got 2.
Sorry, you lose this round.

It's better if you quit now.
```

Let's calculate the probabilty of winning (for bets on the
Pass Line). For a roll of two dice it breaks down like
this:

Sum of dice	2	3	4	5	6	7	8	9	10	11	12
Probability	$\frac{1}{36}$	$\frac{2}{36}$	$\frac{3}{36}$	$\frac{4}{36}$	$\frac{5}{36}$	$\frac{6}{36}$	$\frac{5}{36}$	$\frac{4}{36}$	$\frac{3}{36}$	$\frac{2}{36}$	$\frac{1}{36}$

We can construct the following tree diagram for the process:

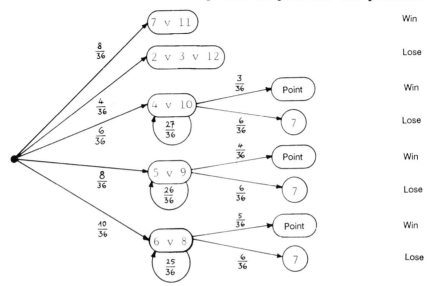

The probability of winning with the first roll is 6/36 + 2/36. If a 4 or a 10 had turned up, the probability of getting the point would have been 1/3. If a 5 or 9 had turned up the probability would have been 2/5. Finally if a 6 or 8 had turned up, it would have been 5/11. Here are the numbers behind the probability of winning:

$$(8/36) + (6/36)*(1/3) + (8/36)*(2/5) + (10/36)*(5/11) =$$
$$(244/495) = 0.49292... \approx 49.3\%$$

The game is almost fair. If you bet 1 dollar at every move, you would win on the average

$$1 * (244/495) + (-1) * (251/495) = -(7/495) = -0.1414...$$

dollars per move. That means that you would lose approximately 1.4 cents per move. The previous tree diagram shows that the average game time (average number of rolls per turn) to be: (557/165) ≈ 3.4

137

Now that all this gambling has encouraged you to take risks with your hard-earned money, let's try a game with coins.

Example 5.3   PENNY TOSS

When a coin is tossed three times there are eight possible combinations that could result. Let's abbreviate HEADS as H and TAILS as T. Here are the combinations that can turn up on the coins: HHH, HHT, HTH, THH, HTT, THT, TTH, TTT. Player A chooses one of these combinations and player B take one of the remaining seven. Then they toss the penny until one of these turns up. The player whose combination appears first wins the game.

Computer simulation is used to answer the question whether players A and B have equal chances or whether B can improve his odds given the fact that A chose first.

Let's develop a program that includes the following dialogue:

```
How many series of tosses do you want?  3
What combination have you chosen: HHT
I'll choose combination: THH
1. Series: THH      MY COMBINATION!
2. Series: HHHHT    YOUR COMBINATION!
3. Series: TTHH     MY COMBINATION!
Results after three tosses:
Your combination came up first 1 times.
My combination came up first 2 times.
```

Here is the program:

```
00100 : PRINT CHR$(147)
00110 : PRINT "        PENNY TOSS
00120 : PRINT "        ----------
00130 :
00140 REM THE OBJECT IS TO ENCOURAGE THE STUDY OF MARKOV
00145 REM CHAINS
```

138

```
00150 :
00160 REM === START =======================================
00170 :
00175 : PRINT : PRINT : PRINT
00180 : PRINT "LET US ASSUME A COIN IS THROWN THREE TIMES.
00190 : PRINT "AMONG THE 8 POSSIBLE COMBINATIONS
00200 : PRINT "HHH, HHT, HTH, HTT, THH, THT, TTH, TTT
00210 : PRINT "YOU CAN CHOOSE ONE COMBINATION.
00220 : PRINT "I SHALL ALSO CHOOSE ONE.
00230 : PRINT "THEN WE TOSS THE COIN (MAKING A SERIES).
00240 : PRINT "THE ONE WHOSE COMBINATION APPEARS FIRST WINS.
00250 : PRINT
00260 :
00270 REM === INITIALIZATION ==============================
00280 :
00290 : FOR I = 1 TO 8 : READ B$(I), C$(I) : NEXT I
00300 :
00305 : PRINT : PRINT
00310 : INPUT "HOW MANY SERIES DO YOU WANT? ";N
00320 : PRINT : INPUT "WHAT IS YOUR COMBINATION  ";B$
00330 :
00340 : FOR I = 1 TO 8
00350 :    IF B$ = B$(I) THEN LET C$ = C$(I) : GOTO 400
00360 : NEXT I
00370 :
00380 : PRINT "WRONG INPUT!" : GOTO 320
00390 :
00400 : PRINT "I CHOOSE THE COMBINATION "; C$
00410 :
00420 : PRINT : PRINT : PRINT "    " CHR$(18) "PRESS ANY KEY
00430 : GET T$ : IF T$ = "" THEN 430
00440 :
00450 DATA HHH, THH, TTT, HTT, HHT, THH, TTH, HTT
00460 DATA HTH, HHT, THT, TTH, HTT, HHT, THH, TTH
00470 :
00480 REM === ROUND =======================================
00490 :
00500 : FOR J = 1 TO N : REM LOOP START--------------------
00510 :
00520 :     PRINT : PRINT J;". SERIES:   ";
00530 :
00540 :     LET W$ = ""
00550 :
00560 :   REM --- COIN TOSS
00570 :
00580 :     IF RND(TI) < 0.5 THEN LET A$ = "H" : GOTO 600
00590 :                         LET A$ = "T"
00600 :     PRINT A$;
00610 :     LET W$ = W$ + A$
00620 :
00630 :   REM --- COMPARE
00640 :
```

```
00650 :      LET M$ = RIGHT$(W$,3) : REM LAST 3 TOSSES
00660 :      IF M$ = B$ THEN 720  : REM PLAYER'S COMBINATION
00665 :                             REM APPEARS
00670 :      IF M$ = C$ THEN 770  : REM COMPUTER'S COMBINA-
00675 :                             REM TION APPEARS
00680 :      GOTO 580               : REM NEW ATTEMPT
00690 :
00700 :   REM --- DETERMINING THE WINNER
00710 :
00720 :      LET B = B+1
00730 :      PRINT "      " CHR$(18) "YOUR COMBINATION!
00740 :      PRINT
00750 :      GOTO 810
00760 :
00770 :      LET C = C+1
00780 :      PRINT "      MY COMBINATION.
00790 :      PRINT
00800 :
00810 :   REM --- WAITING LOOP
00820 :
00830 :      PRINT "       " CHR$(18) "PRESS A KEY.
00840 :      GET T$ : IF T$ = "" THEN 840
00850 :
00860 : NEXT J : REM LOOP END -----------------------------
00870 :
00880 REM === SCORE =======================================
00890 :
00895 : PRINT : PRINT : PRINT
00900 : PRINT "RESULTS AFTER ";N;" SERIES OF TOSSES:
00910 : PRINT
00920 : PRINT "YOUR COMBINATION "; B$;" APPEARED ";B;
00930 : PRINT "TIMES FIRST." : PRINT "MY COMBINATION "; C$;
00935 : PRINT " APPEARED ";C;" TIMES FIRST."
00940 :
```

Explanation: Eight pairs B$(I) , C$(I) from the DATA lines
450, 460 are entered. These express the patterns chosen by
the player and those chosen by the computer. After setting
the number of toss series, the player is asked to enter his
combination B$. Then the computer selects its own combina-
tion C$ (lines 340 - 350).

During the actual game the penny toss is first simulated (Line 580) and the appropriate character H or T is
attached to the already existing character sequence W$. If
W$ equals HHT and the toss turned up H, the result is W$ =
HHTH. By means of

$$M\$ = RIGHT\$(W\$,3)$$

the right three characters are retrieved from the character

sequence W$. These represent the last three tosses. If M$ = B$ then the player's combination has turned up and the series ends; if M$ = C$, then the computer's combination has turned up. Otherwise the penny is tossed again.

When you have played with this program a while you will see that the computer often wins. How does it do it? Rather, how does the machine select its combination of tosses so skillfully?

Let's assume you choose the combination HHH. The probability that it will turn up after the first three tosses is 1/8. All this is assuming that the coins are not double headed or something. If your opponent has chosen a combination craftily he can prevent yours from appearing again first. He just has to be sure that in the fourth toss or later the combination

*HHH   (the * represents the rest of the series)

does not come up first. He can do this by replacing the first H with a T. This is precisely what the computer does. (see DATA line 390 DATA HHH, THH, ...). If a T is tossed before the player's combination comes up, the player cannot win! What happens when you select the combination HTH? The opponent will choose so that the beginning of your combination (HT) appears at the end of his own combination. The computer's choice for the first toss in its combination is H; again, if HH comes up before the player's combination does, the computer must win. What is the probability of its winning?

To answer this question let's draw a so-called transition graph. This gives the possible transitions from one combination to another once the penny has been tossed. This graph

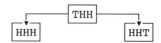

means that after the combination THH has appeared, there is the probability of 1/2 that either the combination HHH or HHT will appear. In other words *THHH or *THHT can follow

141

*THH.   (The three underlined letters is the combination
after the toss.)

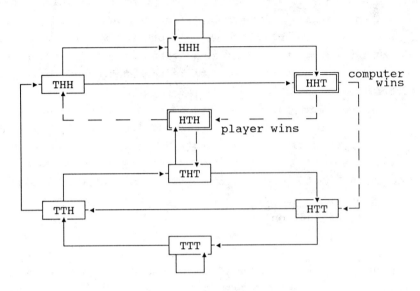

The dotted lines designate <u>impossible</u> transitions.   HHT and
HTH are final results.   Note the symmetry of the figure before the players have chosen their combinations.   All transition probabilities are 1/2.

This graph tells us a lot.   If the game begins with the
combination HHT, THH or HHH then the computer will win.   If
the combination TTH, TTT, or HTT appears first then the combination TTH has to appear before any player can win.   This
way there is a good probability that the computer will win
when the game starts with TTT or HTT, the same as when it
starts with TTH.   Let's use the following abbreviation

$$P(C:HHH)$$

for "the probability that C will win once the combination
HHH has appeared."   Now we can write what we've learned as
follows:

1. $P(C:HHT) = P(C:HHH) = P(C:THH) = 1$

2. $P(C:HTH) = 0$

3. $P(C:TTH) = P(C:TTT) = P(C:HTT)$

We still have to determine the probabilities $P(C:THT)$ and
$P(C:TTH)$.   Once the combination TTH has been tossed then the

next toss will produce either THH or THT with equal probability. Therefore,

4. P(C:TTH) = 1/2 * P(C:THT) + 1/2 * P(C:THH),

and

5. P(C:THT) = 1/2 * P(C:HTT) + 1/2 * P(C:HTH).

Here is the result when we take 1 - 5 into account:

$$P(C:TTH) = 2/3 \quad \text{and} \quad P(C:THT) = 1/3.$$

Following this the probability of winning for C, the computer, is:

$$\frac{1}{8} (1 + 1 + 1 + 0 + \frac{2}{3} + \frac{2}{3} + \frac{2}{3} + \frac{1}{3}) = \frac{2}{3}$$

This means: if you choose combination HTH, the computer will
win two out of three series of tosses. The remaining
choices and probabilities can be derived in corresponding
fashion. In the brainteasers at the end of this chapter you
will find further variations on this interesting game.

                    ****

    Now that you've risked your fortune gambling with dice
and coins, it's time to move on to the card tables. Every-
one has at least heard of the game Blackjack, if not actually played it. It is probably one of the most popular card
games because players stand a chance of realizing some real
earnings. On the other hand it takes immense concentration,
practice and strategic calculation to become adept at Blackjack. Most card players usually find that a full-time job
offers more steady income. The rules for Blackjack can get
complicated and may vary depending on the casino but we've
worked out a modest version of the game that can be played
on the computer.

Example 5.4   TWENTY-ONE

The game uses 32 cards; only their numerical value, neither
their color nor suit, is important. Here are the point values: Jack = 2; Queen = 3; King = 4; 7, 8, 9, 10 are the same
as their printed value; Ace = 11. After the dealer has designated his capital as the bank, the cards are shuffled and
cut. Each player places a bet and then receives two cards
face down. The dealer gets one card face up. Now the play-

er can request more cards. The object is to accrue a point value of 21 or get as close as possible without going over. If a player goes over 21 points, he is out of the game and loses the money he bet. After all the players have drawn cards, the dealer draws if his count is less than 17, and otherwise stays. If the dealer's hand is over 21 he pays all the players still in the game double their entering bets. Otherwise the highest number of points (less than or equal to 21) wins. In the case of a tie, the dealer wins.

Here is a simplified program that concentrates on the essentials of the game. Let's begin by having only one player (you) and the dealer (the computer). Furthermore, we aren't going to place any cash bets but just register if a player has won or lost.

```
Do you want to study the rules? n
The cards are shuffled and cut: Here Goes!
I have drawn: jack.
Your cards: Queen and 9.
Want another card? y
Your card: 8.
Want another card? n
Your total: 20
I have drawn: Queen
I have drawn: Ace
I have drawn: Queen
My total: 19
You win!
How about another round? y
The cards are shuffled and cut: Here Goes!
I have drawn: 8
Your cards: Queen and 10.
Want another card? y
Your card: King
Want another card? y
Your card: 8
Your total: 25.
Sorry. You've lost this round.
How about another round? n
```

```
01000 : PRINT CHR$(147)
01010 : PRINT "          TWENTY-ONE
01020 : PRINT "          ----------
01030 :
01040 REM GAME OF CHANCE USING CARDS; THE COMPUTER HOLDS
01045 REM THE BANK
01050 :
01060 :
01070 REM *** MAIN PROGRAM *******************************
01080 :
01090 : GOSUB 1220 : REM START
01100 :
01110 : GOSUB 1350 : REM INITIALIZATION
01120 :
01130 : GOSUB 1580 : REM ROUND
01140 :
01150 : GOSUB 2370 : REM EXIT OR REPEAT
01160 :
```
145

```
01170 : END
01180 :
01190 REM *** END MAIN PROGRAM ****************************
01200 :
01210 :
01220 REM ### SUBROUTINE 'START' #########################
01230 :
01240 : PRINT : PRINT
01250 : PRINT "DO YOU WANT TO SEE THE RULES? (Y/N)
01260 : GET A$ : IF A$ = "" THEN 1260
01270 : IF A$ <> "Y" THEN RETURN
01280 :
01290 : REM  INSERT THE RULES HERE
01300 :
01310 : FOR I = 1 TO 500 : NEXT I
01320 :
01330 : RETURN : REM END START ########################
01340 :
01350 REM ### SUBROUTINE INITIALIZATION #################
01360 :
01370 : REM === PREPARE SHEET
01380 :
01390 :    DIM B(32) : REM TABLE OF THE DECK OF 32
01400 :
01410 :    LET T = 0
01420 :    FOR I = 1 TO 4
01430 :      FOR J = 1 TO 8
01440 :        LET T = T+1
01450 :        LET B(T) = J
01460 :      NEXT J
01470 :    NEXT I
01480 :
01490 : REM === BEGINNING VALUES
01500 :
01510 :    LET N = 32 : REM NUMBER OF CARDS IN THE PILE
01520 :
01530 :    LET WB = 0 : REM VALUE OF BANK'S CARDS
01540 :    LET WS = 0 : REM VALUE OF PLAYER'S CARDS
01550 :
01560 : RETURN : REM END INITIALIZATION ##################
01570 :
01580 REM ### SUBROUTINE 'ROUND' ########################
01590 :
01595 : PRINT : PRINT
01600 : PRINT "          " CHR$(18) "THE GAME IS STARTING
01610 :
01620 : REM === DEAL CARDS TO THE BANK
01630 :
01640 :    GOSUB 2200 : REM DRAW A CARD
01650 :    PRINT : PRINT : PRINT
01660 :    PRINT "I DREW ";W$
01670 :    LET WB = WB + W
```

```
01680 :
01690 : REM === DEAL CARDS TO PLAYER
01700 :
01710 :    GOSUB 2200 : REM DRAW A CARD
01720 :    PRINT : PRINT "YOU HAVE DRAWN ";W$;" AND ";
01730 :    LET WS = WS + W
01740 :    GOSUB 2200 : REM DRAWING CARD
01750 :    LET WS = WS + W
01760 :    PRINT W$
01770 :
01780 : REM === PLAYER DRAWS CARDS
01790 :
01800 :    PRINT
01810 :    PRINT "CARE TO DRAW A CARD? (Y/N)
01820 :    GET A$ : IF A$ = "" THEN 1820
01830 :    IF A$ <> "Y" THEN 1890
01840 :    GOSUB 2200 : REM DRAWING A CARD
01850 :    PRINT "YOU HAVE DRAWN ";W$
01860 :    LET WS = WS + W
01870 :    GOTO 1800 : REM ANOTHER CARD?
01880 :
01890 : REM === SCORING PLAYER'S CARDS
01900 :
01910 :    PRINT
01920 :    PRINT "THE TOTAL VALUE OF YOUR CARDS IS "; WS
01930 :    IF WS > 21 THEN 2150 : REM PLAYER LOSES
01940 :
01950 : REM === BANK DRAWS CARDS
01960 :
01970 :    PRINT : PRINT
01980 :    GOSUB 2200 : REM DRAWING A CARD
01990 :    PRINT "I HAVE DRAWN ";W$
02000 :    LET WB = WB + W
02010 :    IF WB < 17 THEN 1980 : REM ADDITIONAL CARD
02020 :
02030 : REM === SCORING THE BANK'S CARDS
02040 :
02045 :    PRINT
02050 :    PRINT "THE TOTAL VALUE OF MY CARDS IS: "; WB
02060 :
02070 : REM === GAME DECISION
02080 :
02090 :    IF WB > 21 THEN 2160 : REM PLAYER WINS
02100 :
02110 :    IF WB = 21 THEN 2150 : REM BANK WINS
02120 :
02130 :    IF WB < WS THEN 2160 : REM PLAYER WINS
02140 :
02150 :    PRINT : PRINT : PRINT "I HAVE WON." : GOTO 2180
02160 :    PRINT : PRINT : PRINT "YOU HAVE WON!
02170 :
02180 :    RETURN : REM ### END OF ROUND ##################
```

```
02190 :
02200 : REM +++ SUBROUTINE 'DRAW A CARD' ++++++++++++++++++
02210 :
02220 :    LET R = INT(N*RND(TI)+1)
02230 :    LET C = B(R)
02240 :    LET B(R) = B(N)
02250 :    LET N = N-1 : REM ONE LESS CARD ON THE PILE
02260 :
02270 : REM --- SETTING VALUES
02280 :
02290 :    IF C = 1 THEN LET W$ = "ACE"        : LET W = 11
02300 :    IF C = 2 THEN LET W$ = "JACK"       : LET W = 2
02310 :    IF C = 3 THEN LET W$ = "QUEEN"      : LET W = 3
02320 :    IF C = 4 THEN LET W$ = "KING"       : LET W = 4
02330 :    IF C > 4 THEN LET W$ = STR$(C+2)    : LET W = C+2
02340 :
02350 :    RETURN : REM END OF CARD DRAWING +++++++++++++++++
02360 :
02370 REM ### SUBROUTINE 'EXIT OR REPEAT' ################
02380 :
02390 : PRINT
02400 : PRINT "ANOTHER ROUND? (Y/N)
02410 : GET A$ : IF A$ = "" THEN 2410
02420 : IF A$ = "Y" THEN RUN
02430 :
02440 : RETURN : REM ### END OF 'EXIT OR REPEAT' ##########
02450 :
```

The most famous game of chance is of course Roulette. This
wheel of fortune is probably the most well known symbol of

gambling that exists. There is no other game in which losing and winning follow so closely on the heels of one another. No other game of chance has so affected the literary
imagination: think of Dostoyevsky's story "The Gambler" or
Andre Malraux's novel La Condition Humaine. It was thought
that the philosopher Blaise Pascal (1623-1662) invented the
game. This has been proven to be untrue although Pascal did
write several treatises called "La Roulette." It seems he
was refering to a curve, nowadays called the "cycloid,"
which is the shape made by a point on the rim of a rolling
wheel. At least we know that the game we now call roulette
was introduced in Paris in 1765 by Gabriel de Sartine though
similar versions were known earlier.

It's a good idea to become thoroughly acquainted with
the rules of the game and the chances of winning before you
take your inheritance and play for real. Of course you
could buy a toy roulette wheel and figure out the game
through sheer practice. But a better way is to simulate the
game on the computer -- that fickle game player that has
already made and lost so many fortunes.

Jansenist Roulette Game   (18th Century)

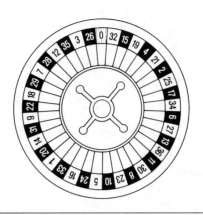

Example 5.5 ROULETTE

The roulette wheel is a hand turned "tub" sunk in a table
and mounted on bearings so as to spin easily. The moveable
part is metal but it is mounted in wood at one end of the
roulette table. To move the wheel, the croupier turns the
cross-shaped handle on the top of the wheel. Around the
edge of the wheel the numbers from 0 to 36 are arranged in a
particular order. The croupier throws a little ball around
the edge of the turning wheel. This then falls into one of
the small felt-lined compartments below a number.

The roulette table is covered with green felt and has
numbers and fields on it that tell the possible bets a player may place. Here they are:

Single number (plein). Player wins 35 times his bet.

Two numbers (cheval). Player wins 14 times the bet.

Three numbers (transversale plein). Player wins 11 times
the bet.

Four numbers (carre). Player wins 8 times the bet.

Six numbers (transversale simple). Player wins 5 times the
bet.

Dozens (1-12, 13-24, 25-36). Player wins double the bet.

Columns. Player wins double the bet.

Simple odds. These are Manque, Passe, Even, Odd, Red,
Black). Player wins the amount of his bet.

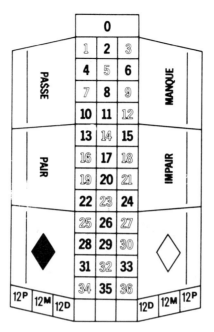

If the ball lands in zero, then all the "simple chances"
are blocked. The croupier pushes the chips to the blocking
line. A player can now ask that his bet be divided. In
this case half the bet goes to the bank and the player gets
the other half back. If a player foregoes dividing his bet
than his bet is freed and pulled back from the blocking line
if the number he had bet on comes up next. If zero comes up
next then all blocked bets become the property of the bank.

Our program simulates a roulette game with only one player
against the bank. We shall omit the complicated rules that
apply to zero. Also -- to simplify the program and make the
stakes higher -- we are leaving out the bets 2, 3, 4, and 5.

Dialogue:

151

```
You may place the following bets:
Number (plein)....................1 through 36
First Dozen.................................37
Second Dozen...............................38
Third Dozen................................39
(..........................................)
Red (rouge)................................47
Black (noir)...............................48
What kind of bet are you placing?  15
How much are you betting? 20
Any more bets? y
What kind of bet are you placing? 47
How much are you betting? 30
Any more bets? n
                RIEN NA VAS PLUS
The ball has landed in 14.
With your first bet (15 plein) you lost $20.00.
With your second bet (rouge) you won $30.00.
```

Here is the program:

```
01000 : PRINT CHR$(147)
01010 : PRINT "        ROULETTE
01020 : PRINT "        --------
01030 :
01040 REM SIMULATION OF THE CASINO GAMBLING GAME FOR ONE
01045 REM PERSON
01050 :
01060 REM === INITIALIZATION ==============================
01070 :
01080 : REM VARIABLES:
01090 : REM    N . . . . . . . . . . . NUMBER OF BETS
01100 : REM    I . . . . . . . . . . . NUMBER OF THE BET
01110 : REM    T, T(I) . . . . . . . TYPE OF BET NO. I
01120 : REM    E, E(I) . . . . . . . BET NO. 1
01130 : REM    A . . . . . . . . . . PAYMENT/LOSS
01140 :
01150 : LET M = 5              : REM MINIMUM BET
01160 : LET H = 1000           : REM MAXIMUM BET
01170 :
01180 : LET CB = 10000         : REM BANK'S CAPITAL
01190 :
01200 : PRINT : PRINT "HOW MUCH MONEY DO YOU HAVE? ";
01210 : INPUT CS
01220 : IF CS < M THEN PRINT "NOT ENOUGH!" : GOTO 1200
01230 :
```

```
01240 REM === PLACE YOUR BETS ==============================
01250 :
01260 : LET I = 0      : REM NUMBER OF THE BET
01270 : LET CL = CS    : REM PLAYER'S CAPITAL
01280 :
01290 : LET I = I + 1
01300 :
01310 :    PRINT
01320 :    PRINT "YOU CAN PLACE THE FOLLOWING BETS:
01330 :    PRINT
01340 :    PRINT "NUMBER (PLEIN) . . . . . . . . . . 1 TO 36
01350 :    PRINT "FIRST DOZEN. . . . . . . . . . . . . .37
01360 :    PRINT "SECOND DOZEN . . . . . . . . . . . . .38
01370 :    PRINT "THIRD DOZEN. . . . . . . . . . . . . .39
01380 :    PRINT "FIRST COLUMN . . . . . . . . . . . . .40
01390 :    PRINT "SECOND COLUMN. . . . . . . . . . . . .41
01400 :    PRINT "THIRD COLUMN . . . . . . . . . . . . .42
01410 :    PRINT "1-18 (MANQUE). . . . . . . . . . . . .43
01420 :    PRINT "19-36 (PASSE). . . . . . . . . . . . .44
01430 :    PRINT "EVEN (PAIR). . . . . . . . . . . . . .45
01440 :    PRINT "ODD (IMPAIR) . . . . . . . . . . . . .46
01450 :    PRINT "RED (ROUGE). . . . . . . . . . . . . .47
01460 :    PRINT "BLACK (NOIR) . . . . . . . . . . . . .48
01470 :
01480 :    PRINT : PRINT : PRINT "CHOOSE ";
01490 :    INPUT "TYPE OF BET "; T
01500 :    IF T < 1 OR T > 48 OR T <> INT(T) THEN 1480
01510 :    LET T(I) = T
01520 :
01530 :    INPUT "HOW MUCH ARE YOU BETTING? "; E
01540 :    IF E < M THEN PRINT "LEAST BET IS "; M : GOTO 1530
01550 :    IF E > H THEN PRINT "MOST BET IS "; H : GOTO 1530
01560 :    IF E > CL THEN PRINT "NOT SO MUCH !" : GOTO 1530
01570 :    LET E(I) = E
01580 :
01590 :    PRINT : PRINT : PRINT "ANOTHER BET? (Y/N)
01600 :    GET A$ : IF A$ = "" THEN 1600
01610 :
01620 :    IF A$ = "Y" THEN CL = CL-E : GOTO 1290
01625 :                              REM ANOTHER BET
01630 :
01640 REM === THE WHEEL IS SPINNING =======================
01650 :
01660 : PRINT : PRINT "        " CHR$(18) "RIEN NE VAS PLUS
01670 : PRINT "        THE WHEEL OF FORTUNE TURNS
01680 : FOR J = 1 TO 1000 : NEXT J : REM WAITING LOOP
01690 :
01700 : LET C = INT(37*RND(TI)) : REM RANDOM NUMBER BETWEEN
01705 :                              REM 0 AND 36
01710 :
01715 : PRINT : PRINT
01720 : PRINT "THE BALL HAS LANDED IN "; CHR$(18) C
```

153

```
01730 :
01740 REM === DETERMINING WINS AND LOSSES  =================
01750 :
01760 : LET N = I : REM NUMBER OF BETS PLACED
01770 :
01780 : PRINT : PRINT : PRINT
01790 :
01800 : FOR I = 1 TO N
01810 :
01820 :    PRINT "WITH BET NUMBER";I;
01830 :
01840 :    LET E = E(I)
01850 :
01860 :    IF C = 0 THEN LET A = -3 : GOTO 1940
01870 :
01875 :    LET X = T(I)
01880 :    IF X <= 36 THEN GOSUB 2150 : GOTO 1940
01890 :    ON INT((X-37)/4) + 1 GOTO 1900,1905,1910
01900 :    ON X-36 GOSUB 2210, 2270, 2330, 2390 : GOTO 1940
01905 :    ON X-40 GOSUB 2450, 2510, 2570, 2630 : GOTO 1940
01910 :    ON X-44 GOSUB 2690, 2750, 2810, 2880 : GOTO 1940
01920 :    PRINT "DOES NOT FUNCTION"
01930 :
01940 :    IF A > 0 THEN PRINT  A;" DOLLARS WON.": GOTO 1960
01950 :                  PRINT -A;" DOLLARS LOST.
01960 :    LET CS = CS + A : REM PLAYER'S NEW CAPITAL
01970 :    LET CB = CB - A : REM BANK'S NEW CAPITAL
01980 :
01990 : NEXT I : REM EVALUATION OF NEXT BET
02000 :
02010 : PRINT
02020 : PRINT "YOU NOW HAVE "; CS; " DOLLARS.
02030 :
02040 : IF CS < M THEN PRINT "IT'S TIME TO QUIT NOW." : END
02050 : IF CS > CB THEN PRINT "THE BANK IS BROKEN!!! " : END
02060 :
02070 : PRINT : PRINT : PRINT : PRINT "ANOTHER GAME? (Y/N)
02080 : GET A$ : IF A$ = "" THEN 2080
02090 : IF A$ = "Y" THEN 1240
02100 :
02110 : END
02120 :
02130 REM ### SUBROUTINES TO EVALUATE THE BETS
02140 :
02150 :    REM +++ NUMBER +++
02160 :
02170 :       PRINT "(";T(I);" PLEIN) ";
02180 :       IF C = T(I) THEN LET A = 35*E : RETURN
02190 :                        LET A = -E   : RETURN
02200 :
02210 :    REM +++ FIRST DOZEN +++
02220 :
```

```
02230 :      PRINT "(12P)       ";
02240 :      IF 1 <= C AND C <= 12 THEN LET A = 2*E : RETURN
02250 :                            LET A = -E  : RETURN
02260 :
02270 :  REM +++ SECOND DOZEN +++
02280 :
02290 :      PRINT "(12M)       ";
02300 :      IF 13 <= C AND C <= 24 THEN LET A = 2*E : RETURN
02310 :                            LET A = -E  : RETURN
02320 :
02330 :  REM +++ THIRD DOZEN +++
02340 :
02350 :      PRINT "(12D)       ";
02360 :      IF 25 <= C AND C <= 36 THEN LET A = 2*E : RETURN
02370 :                            LET A = -E  : RETURN
02380 :
02390 :  REM +++ FIRST COLUMN +++
02400 :
02410 :      PRINT "(1ST COLUMN) ";
02420 :      IF C - 3*INT(C/3) = 1 THEN LET A = 2*E : RETURN
02430 :                            LET A = -E  : RETURN
02440 :
02450 :  REM +++ SECOND COLUMN +++
02460 :
02470 :      PRINT "(2ND COLUMN) ";
02480 :      IF C - 3*INT(C/3) = 2 THEN LET A = 2*E : RETURN
02490 :                            LET A = -E  : RETURN
02500 :
02510 :  REM +++ THIRD COLUMN +++
02520 :
02530 :      PRINT "(3RD COLUMN) ";
02540 :      IF C - 3*INT(C/3) = 0 THEN LET A = 2*E : RETURN
02550 :                            LET A = -E  : RETURN
02560 :
02570 :  REM +++ MANQUE +++
02580 :
02590 :      PRINT "(MANQUE)     ";
02600 :      IF 1 <= C AND C <= 18 THEN LET A =  E : RETURN
02610 :                            LET A = -E : RETURN
02620 :
02630 :  REM +++ PASSE +++
02640 :
02650 :      PRINT "(PASSE)      ";
02660 :      IF 19 <= C AND C <= 36 THEN LET A =  E : RETURN
02670 :                            LET A = -E : RETURN
02680 :
02690 :  REM +++ EVEN +++
02700 :
02710 :      PRINT "(PAIR/EVEN)       ";
02720 :      IF C - 2*INT(C/2) = 0 THEN LET A =  E : RETURN
02730 :                            LET A = -E : RETURN
02740 :
```

```
02750 :    REM +++ ODD +++
02760 :
02770 :       PRINT "(IMPAIR/ODD)          ";
02780 :       IF C - 2*INT(C/2) = 1 THEN LET A =   E : RETURN
02790 :                                   LET A = -E : RETURN
02800 :
02810 :    REM +++ RED +++
02820 :
02830 :       PRINT "(ROUGE/RED)          ";
02840 :       GOSUB 2950
02850 :       IF   R   THEN LET A =  E : RETURN
02860 :                     LET A = -E : RETURN
02870 :
02880 :    REM +++ BLACK +++
02890 :
02900 :       PRINT "(NOIR/BLACK)             ";
02910 :       GOSUB 2950
02920 :       IF NOT R THEN LET A = -E : RETURN
02930 :                     LET A = -E : RETURN
02940 :
02950 :    REM +++ SUBROUTINE 'RED' +++
02960 :
02970 :       LET R1 = (C=1  OR C=3  OR C=5  OR C=7  OR C=9)
02980 :       LET R2 = (C=12 OR C=14 OR C=16 OR C=18 OR C=19)
02990 :       LET R3 = (C=21 OR C=23 OR C=25 OR C=27 OR C=30)
02995 :       LET R4 = (C=32 OR C=34 OR C=36)
03000 :       LET R  = R1 OR R2 OR R3 OR R4
03010 :       RETURN
```

Let's end this chapter with a game based on the nefarious slot machine that has given Las Vegas its reputation: the One-Armed Bandit. By the way, did you know that in Vegas there is one slot machine for every inhabitant of the city?

Example 5.6 ONE-ARMED BANDIT
This coin gulping robot sets three wheels in motion that display symbols such as bells, grapes, oranges, bars, etc. The combination that appears in the windows determines how many coins spew out to you. The object is to write a program that simulates this machine.

Dialogue:

```
Insert a coin.
      The wheels are turning.

              cherry
              bell
              bell
Sorry. You didn't win anything.
Another game? y
Insert a coin.
      The wheels are turning.

              orange
              orange
              orange
Not bad!  You have won 2 dollars.
Another game? n
```

Here is the payoff table showing what the machine will pay for certain combinations.

PAYOFF TABLE	
COMBINATION	PAYOFF (NICKELS)
200
100
18
18
14
14
10
10
5
2

Our program will only have three symbols: bell, orange and cherry. You win if the same three symbols appear in all three places. For three bells, the machine pays $3.00; for three oranges, it pays $2.00, and for three cherries, $1.00. The coin you insert each time is a nickel.

In the following program we have had to simplify a little bit. A margin for error appears when we round off the first nickel. Try the program omitting line 590.

```
00100 : PRINT CHR$(147)
00110 : PRINT "       ONE-ARMED BANDIT
00120 : PRINT "       ----------------
00130 :
00140 REM SIMULATION OF A SLOT MACHINE
00150 :
00160 REM === START ========================================
00170 :
00180 : PRINT
00190 : PRINT "YOU PUT A COIN IN AND PULL DOWN THE LEVER ON
00195 : PRINT " THE RIGHT.
00200 : PRINT "IF 3 BELLS SHOW,    YOU WIN 3 DOLLARS.
```

```
00210 : PRINT "IF 3 ORANGES SHOW,  YOU WIN 2 DOLLARS.
00220 : PRINT "IF 3 CHERRIES SHOW, YOU WIN 1 DOLLAR.
00230 :
00240 : PRINT
00250 : INPUT "HOW MUCH CAPITAL DO YOU HAVE "; CA
00260 :
00270 : PRINT "INSERT A COIN.
00280 :
00290 REM === THE WHEELS ARE SPINNING =====================
00300 :
00310 : PRINT
00320 : PRINT "        " CHR$(18) "THE WHEELS ARE SPINNING
00330 : FOR I = 1 TO 1000 : NEXT I
00340 :
00350 : LET BE = 0 : LET OG = 0 : LET CH = 0
00360 :
00370 : FOR J = 1 TO 3
00380 :
00390 :    LET Z = INT(3*RND(TI)+1)
00400 :    IF Z=1 THEN PRINT "BELL" : LET BE=BE+1 : GOTO 440
00410 :    IF Z=2 THEN PRINT "ORANGE" :LET OG=OG+1 : GOTO 440
00420 :               PRINT "CHERRY" :LET CH=CH+1
00430 :
00440 : NEXT J
00450 :
00460 REM === RESULTS ==================================
00470 :
00480 : PRINT
00490 : PRINT "YOU WON ";
00500 :
00510 : IF BE = 3 THEN PRINT "$3" : LET CA = CA+3 : GOTO 560
00520 : IF OG = 3 THEN PRINT "$2" : LET CA = CA+2 : GOTO 560
00530 : IF CH = 3 THEN PRINT "$1" : LET CA = CA+1 : GOTO 560
00540 : PRINT "NOTHING. TOO BAD."
00550 :
00560 REM === ANOTHER GAME =================================
00570 :
00580 : LET CA = CA - 0.05
00590 : LET CA = INT(100*CA + 0.5)/100
00600 :
00610 : PRINT
00620 : PRINT "YOUR CAPITAL NOW AMOUNTS TO ";CA;"DOLLARS.
00630 :
00635 : PRINT : PRINT : PRINT : PRINT : PRINT
00640 : PRINT "ANOTHER GAME? (Y/N)
00650 : GET A$ : IF A$ = "" THEN 650
00660 : IF A$ = "Y" THEN 270 : REM NEW GAME
00670 :
00680 : PRINT "A WISE DECISION. GOOD-BYE."
00690 :
00700 : END
```

BRAINTEASERS

1. Expand the game Six Hex into a game for N players. Let the computer be the game master.

2. Devise a game and program for other "Hex Numbers." For example, try a game of "One Hex." In each case let the player choose the "hex number".

3. Expand the game "Six Hex" so that you can throw k number of dice instead of only two each turn. Show that the optimal strategy looks like this: a player keeps throwing if the point total reached is smaller than:

$$3k(5/6)^k / (1 - (5/6)^k)$$

(not including the final throw).

4. ODDS - EVENS One of the oldest known dice games has the following rules: a player throws the die as often as necessary before turning up an odd number. Place bets on how many rolls you can throw before this happens. Write a program for this game.

5. How great is the probability that the number series 1, 2, 3, 4, 5, 6, will appear when thirteen good dice are thrown all at once? Answer this problem with the help of a simulation and some mathematics. Hint: here is the solution:

$$1 - \frac{6*5^{13} - 15*4^{13} + 20*3^{13} - 15*2^{13} + 6}{6^{13}}$$

6. ST. PETERSBURG The mathematician Nicolas Bernoulli invented a game called the Saint Petersburg Game. It goes like this: a die is rolled as often as possible until a 6 turns up. If this happens after n rolls, the player is paid n units of money. How much money does one get paid in the long run and how high does the ante have to be (how many

units?) in order for the game to be fair? Solve this one using simulation and a little math.

7. DICE ON ICE Five dice are thrown together. After every throw one or more is removed and "put on ice." Keep playing with the others until all have been put on ice. The goal is to maximize the point value of the dice put on ice. If s is the point value (meaning the number of dots) that turn up then the player gets from every opponent s - 24 points. If s < 24 then the player has to give each opponent 24 - s points. Develop a program with optimal game strategy.

8. ALL SIXES Five dice are thrown. Those showing a six are then removed and laid aside. Keep playing with the others until all the dice have been disposed of in the same way. How long does the game last on the average before this situation is reached?

9. TWENTY-THREE Four dice are thrown simultaneously. The object is to use the four basic arithmetic operations to combine the four point values displayed to get the number 23 (or lower, but as close to 23 as possible). Each number must be used exactly once. Here is an example (for a throw of 1,4,5,6):
    a) (6/1) * 4 - 5 = 24 - 5 = 19
    b) 4*6 - (5-1) = 24 - 4 = 20
    c) 5 * (6-1) - 4 = 25 - 4 = 21
    d) 4 * (5-1) + 6 = 16 + 6 = 22
    e) (4-1) * 6 + 5 = 18 + 5 = 23
    f) (6+1) * 4 - 5 = 28 - 5 = 23
Write a program for this game.

10. TARGET 27 Two players take turns throwing the dice. The face value of the dots that turn up can either be counted toward the player's total or subtracted from the opponent's total. The player who first gets a total between 26 and 29 is the winner. Write a program for this game with the computer as opponent.

11. GERMAN POKER   The game uses six dice; each player gets three throws per round.   A player can decide which dice he will let stand.   A player's points are tallied up after the third throw.   The first to reach 10,000 points wins.   A 1 counts 100 points, a 5 counts 50 points.   All the others alone count nothing.   But rolling 6 three times = 600 points; 5 three times = 500 points; 4 three times = 400 points; 3 three times = 300 points; 2 three times = 200 points; and 1 three times = 1,000 points.

12. CIRCUS DICE ROLL   Sam Lloyd describes the following game: a table is divided into 6 squares, numbered from 1 to 6.   Players can place as much money as they want in these squares.   Three dice are rolled.   If one of the dice turns up the number of the square where you have placed your money, then you get not only the cash you put down but also a bonus of the same amount.   If two dice show the number of your square then you get not only your own money back but also a bonus of double the amount.   If your square number shows up on all three of the dice, you're really in luck: you get your cash plus three times that total.   Here's the catch: if your square doesn't show up on any of the dice, then the bank rakes in your cash.   Who stands to win in this game, anyway?   To find out, write a simulation program or work it out mathematically.

13. Write a simulation program for the game of Crap Shoot that will confirm the values of the theoretically derived winning probabilities and the average game length.

14. Let's assume that a player begins the Crap Shoot with $10.00 and bets $1.00 per move until he has lost all his capital.   Invent a simulation that proves a player can bet

$$10 * (495/7) \approx 707$$

dollars before facing financial ruin.   Your program should contain an array of the following format:

Number	Number of moves to ruin	Number of throws	Maximum attainable income
Average value	707	2390	35

15. Our Crap Shoot simulated a game of one player against the bank. Now write a program for several players, one of whom must be the banker (or "shooter"). Roll the dice to decide who does this job; the highest number designates the shooter. Or, auction off the position of first banker to the highest bidder. He then bets some amount that he will win the game. The other players bet against him and their total bets must add up to the amount of the shooter. Now the shooter rolls. If he wins with a natural then he rakes in all the money and -- if he wants -- is the shooter again in the next round. If he rolls a crap and loses, then he must pay each player double that player's bet from the bank. He can -- if he wants -- still be the shooter. If he loses with a 7 trying for his "point"" then he must pay up as before and surrender the dice.

16. Run a simulation to predict the winning probabilities when you bet on "big 6" or "big 8" (i.e., when 6 or 8 turn up before 7). Here's the equation:

$$(5/11) \approx 45.5\%$$

17. Every gambling game has many variations when it comes to betting systems. This applies to Crap Shoot as well. The system of elimination goes like this: write the numbers 1, 2, 3, 4, 5, 6, 7, 8, 9, 10 on a piece of paper. Then add the first and last numbers in the series and bet the total -- in the first case 11. If you win, then these numbers are eliminated from your series leaving 2, 3, 4, 5, 6, 7, 8, 9. But after you lose a round the amount of the bet gets added to your series: 1, 2, 3, 4, 5, 6, 7, 8, 9, 10, 11. Now continue making your bets the total of the first and last num-

bers in the series. The theory behind this tactic is this: because there are almost the same number of wins as losses during the course of a game, in the long run the entire list will be eliminated. This means that a player will have won the sum total of the numbers in the original series. Run a simulation program to show that the system doesn't ultimately lead to sucess. Where does the false assumption lie?

18. Rewrite the program for Penny Toss so that it can estimate the probabilities of winning depending upon the computer's choice of combination. Think about this: who wins with the combination HTH and THT -- and with what probability?

19. Check the DATA lines 450 and 460 of the program for Penny Toss to see if the winning strategy also works for counterfeit coins. These are coins for which the probability of "heads" is not 1/2.

20. Rewrite the program for Penny Toss so that the computer selects its combination at random and the player chooses afterwards.

21. Generalize the game Penny Toss to combinations involving four tosses, for example HHHH, HHHT, HHTH, ...

22. Rewrite the Penny Toss program for three players. Does the previous strategy remain optimal?

23. Four coins are lying on the table, TAILS facing up. Throw each coin as many times as necessary until HEADS is showing. How many throws are necessary on the average to have them all turned over?

24. Expand the program Twenty-one so that several players can participate. Add a betting component to your new version.

25. In Twenty-one give the player the possibility of letting the Ace count either 1 point or 11 points.

26. Expand the program for Roulette so that you can also bet on "Cheval," on the transversals, and on "Carre."

27. Make it possible for several players to play Roulette.

28. Build into the program for Roulette the possibility of splitting the bet evenly when zero appears.

29. The techniques for entering information into the Roulette program should be improved. For example, make it impossible for a player to accidentally make the same bet twice.

30. The Roulette program is the perfect mechanism for testing the various betting systems (and to discover that they are worthless).
-- The best known is the so-called d'Alembert system. A player bets on the simple chances and raises his bet a fixed amount after every loss. After each win he decreases his bet by this amount.
-- The Martingale system is one that doubles the bet after every loss (possibly even adding 1 as well). Think about why this system also fails.
-- In example 17 we've already had a look at the system that uses the process of elimination.
-- Then there is the Cuban system. This is based on the fact that the third column (3, 6, 9, ...36) has 8 red and only 4 black numbers. Try betting on black and on column 3 at the same time. (Then think about why the system is based on a false assumption). Each of these systems is an algorithm and can easly be incorporated into a program.

31. Expand the program for One-armed Bandit to include the other paying combinations.

32. The program for One-armed Bandit is very generous to the player who reaps big winnings in the game. How should the betting be regulated so that the computer can at least return a little profit to its owner? That guy keeps losing his shirt to computer whiz kids who have studied this book.

33. 7-UP   Draw this figure on a piece of paper.

7	
2	3
4	5
6	8
9	10
11	12

Players can bet on any number. If the banker throws a number from the right hand column, then he gets all the bets on this side but has to double all the bets on the left side. Correspondingly, when he throws a number appearing in the left column, then he does the same for bets on the right side. If he throws a 7 then he wins all the bets in the right and left columns but has to pay off triple those bets on the 7. Program this game.

34. Variations on 7-UP   The banker pays out double the bets of the number that comes up. He pays the bets of the numbers in the same row only at their cash value. 7 wins triple the amount bet. He rakes in the rest of the bets for the bank.

35. Use both computer simulation and mathematical calculations to show that the banker of the game 7-UP has the advantage over a long time. Furthermore, whoever bets on 7 loses one third of his capital in the long run. (This is for variation number 34 above). How does this work for number 33?

36. SHOOT OUT   Three bad guys (each at
one corner of a triangle) are
having a shoot out.

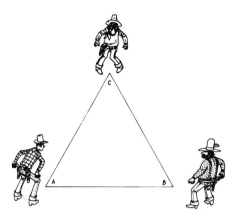

The probabilities of anybody getting hit is 1, 0.8, 0.5 for
the pattern A, B, C. Then they fire their six-guns around
the triangle in that order, dead bandits of course missing
their turns, until only one is left alive. Each player can
choose his target at will. What are their chances of survival if they shoot with optimal strategy? Is there a les-
son in this 3-way duel that might be applied to internation-
al relations?

37. We start by putting one red ball into each of 12 containers. The banker then adds to each container a certain
number (it can also be zero) of white balls. Now the positions of the containers are switched around and a player
chooses one of them at random and takes out a ball. If the
ball is red then the player wins double his bet, but if it
is white, he loses it all. How many white balls should the
banker add to the containers so that he will keep the advantage in the long run <u>and</u> can simultaneously keep the game
attractive to players? The banker's advantage shouldn't be
too great. Write a program to simulate the game.

38. MAGIC DICE   Rolf Mueller writes in TETRA: "I admit it:
this business isn't exactly fair but astonishing and appeal-
ing nonetheless. Take four dice and renumber them. The

first die should have the number 3 on all six sides. The
second die gets 2 on four sides and 6 on the other two. The
third die should have 1 on three sides and 5 on the other
three. On the fourth one put 4 on four sides and 0 on two.

Now all you need is a victim for the following game:
let your partner select one die and you then take another.
Both of you roll your dice simultaneously. The player whose
die shows the higher face value gets one plus point. After
a while your point total will start to outstrip that of your
opponent so that you will have about twice as high a score.
Your opponent is going to get itchy, maybe even dangerous.
He demands better playing equipment, mayber he even wants
your die. Go ahead and give it to him and choose one of the
remaining three dice and continue the game with both of you
in a better frame of mind. But your luck soon becomes apparent again and as your point total mounts, so does your
opponent's frustration. He's going to want the die you are
using, thinking it's better. Be generous and part with it.
Now in the third round it turns out that you are still twice
as good as your opponent with your new choice of die.

It now looks as though the best of all possible dice
has been discovered -- and it is naturally in your hands!
If you part with this one as well, this best of all possible
dice, he still won't be able to hold a candle to your incredible score. Why? His curse is that he picks first from
the dice. Your subsequent choice was always a better one.

Write a program that gives the computer the counter-
choice and thereby the opportunity to win in the long run.

39. A variation on Bradly Efron -- Mueller's game: let die A
have the numbers 1, 2, 3, 3, 6, 6; and let die B have the
numbers 1, 1, 2, 5, 6, 6. Show that A will win against B
with the statistical probability of 16/36.

Let die C have the numbers 2, 2, 4, 4, 4, 5. Then B
will win over C with the probability of 17/36 but C will
beat A with the probability of 18/36!

Confirm this using computer simulation as well as mathematical calculations. Invent more "magic" dice.

"Frankly, I liked the old games better!"

CHAPTER VI

GAMES OF STRATEGY

> Nowhere has man exhibited more
> ingenuity than in his games.
>
> Leibnitz

In this chapter we are going to explore games that all have certain similarities:

a) They are contests between two players.

b) They terminate after a finite number of moves.

c) There are no elements of chance involved (no dice are used or cards shuffled). The result is that the game is determined solely by the players' decisions.

d) Both players can freely observe each other's moves and evaluate the changes that these make in the game.

The terminology is already pretty familiar: a series of game positions is called a "round"; the player's individual decision within a round is a "move"; a player's "strategy" is a plan involving decisions he makes when it is his move. The "strategy" involved in one move determines what move shall logically follow when it is that player's move again.

****

The game called Nim (with its different variations) is one of the simplest games of strategy that has been completely analyzed and programmed for the computer without being trivialized. In 1612 Bachet de Meziriac wrote a work "Problemes plaisants et delectable, qui se font par les nombres" (Pleasant and Entertaining Games with Numbers), in which he described a game known today as Pick-up-Sticks. This has been translated into a number game and involves, basically, games in which players use strategy when removing items from play. At around the turn of this century American college students played the game, calling it Fan Tan. The name Nim comes from the Canadian mathematician Charles L. Bouton who analyzed it in 1902.

Example 6.1 NIM WITH SEVERAL BOWLS

On a table there are three bowls of fruit. Two players take turns removing as much fruit as they like from any one of the bowls (but at least one piece of fruit per move). The player who can no longer remove a piece, because all the bowls are empty, loses. The computer is one of the players.

It is important to teach the computer a strategy so it will be an opponent that can be taken seriously in this game. Let's take a situation in which one of the bowls is already empty and play the round to the end. It is player A's turn.

Game status	Move
✳ ✳ ✳ ✳ ✳ ✳	A takes 2 pieces from bowl 2
✳ ✳ ✳ ✳	B takes 1 piece from bowl 1
✳ ✳ ✳	A takes 1 piece from bowl 2
✳ ✳	B takes 1 piece from bowl 2
✳	A takes 1 piece from bowl 1
	All the bowls are empty; B loses!

What strategy did A use? Each time he took away enough pieces so as to leave the same number of pieces in each of the two bowls. Whatever B could have done, player A could always restore the balanced situation ("same number of pieces in each bowl"). In doing so, A was demonstrating a winning strategy. (If there had been the same amount of fruit in each bowl then B, who moved next, would have had a winning strategy, because A would have had to upset the equal distribution of the fruit with his first move. In this case B would have proceeded as A did in the example.)

We can describe the "game status" as the number of pieces of fruit still in the bowls. When we are talking about three bowls, we can abbreviate the situation with the following three numbers (2,4,0) which tells us that every position

$$(n,m,0) \text{ with } n \neq m$$

is a winning position for the player whose move it is. Every position

$$(n,n,0)$$

is a losing position for the player about to move. At least he will lose the round if the opponent is in the know about the strategy.

How does it look when every bowl has fruit in it? Let's take this case (3,5,7):

Is there a move that can produce the "balanced" situation -- a situation an opponent must destroy but which can be restored again in the following move? What do we now mean by "balanced"? It cannot mean "same number of pieces in each bowl"; that would mean a move which is against the rules. How about the following possibility:

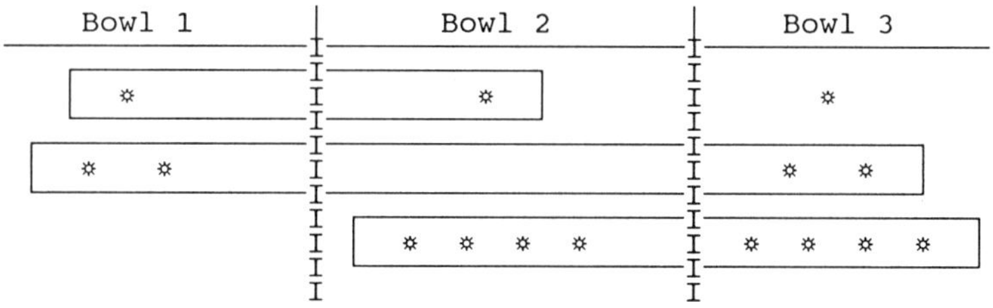

Thus, we remove one piece from bowl 3 and create the game status of (3,5,6). Our opponent will now have to destroy the balance no matter what his move may be. Let's assume he takes 5 pieces from bowl 3 and creates the status (3,5,1). The balance is now completely destroyed and we restore it in the next move like this:

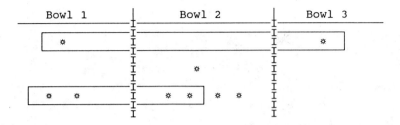

| Bowl 1 | Bowl 2 | Bowl 3 |

By taking three pieces from bowl 2 we create the game status
(3,2,1). This is a losing position for our opponent. What
can he do? Try playing through the various possibilities.

We can make the following generalizations about the
possible game positions. Every game position (created by
the distribution of fruit in the three bowls) is either a
winning position or a losing position. The following are
true:

1. The final position (0,0,0) is a losing position.

2. A position is a winning position if it forces a following
move that results in a losing position.

3. A position is a losing position if every possible move
makes it into a winning position.

This should clarify the strategy: <u>a player must try to present the opponent with a losing position.</u> Then the opponent
will be forced (according to point 3, above) to leave behind
a winning position after he has moved. This means: if the
beginning position is a winning position the player whose
move it is always wins if he follows this strategy. When it
is his move he must find a strategy that will force his
opponent to make a move that will result in a losing position (point 2 above). Herein lies the art of playing well.
Since the number of pieces of fruit decreases with every
move, the game ends after a finite number of moves with the
position (0,0,0).

So far, so good? Okay, then -- how do you recognize
winning or losing positions? <u>One</u> possibility is this:
starting with the losing position (0,0,0), we write down all
positions that can be changed into (0,0,0) in one move and
call these winning positions. These positions we call "predecessors." Starting with every winning position found in
this way, we then search for all predecessors and mark those

174

among them that lead to winning positions as losing positions; the rest we mark winning positions. This process is repeated until all positions are marked. (It can also be done using the computer!)

A _second_ way to recognize the different positions follows from the preceding strategy. We had divided the number of pieces of fruit into powers of 2 and then identified the "balanced" positions (i.e., the losing positions) as those where the number of powers of 2 was even. (In the case (3,5,6) the result is 3=2+1, 5=4+1 and 6=4+2. In other words, the number of 4's, 2's and 1's was even). Or, to put it more formally: the amounts of pieces of fruit are written in binary code and then their Nim total is calculated (0+0=0, 0+1=1+0=1, 1+1=0 using pointwise addition). If the total is zero, you have a losing position, otherwise it is a winning position.

In the example above it looks like this: 3 = 11
5 = 101
6 = 110
_____
000, which
means that (3,5,6) is a losing position. It is easy to show that the winning and losing positions fulfill the criteria of points 1, 2, and 3 above. This is Bouton's famous method and it is our job to translate it into a computer program. Our program will have the following structure:

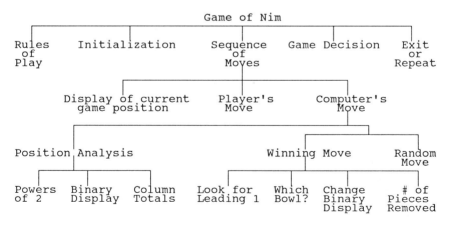

In writing this program we must make sure that the subroutines don't get strewn around the entire program and that they respect the hierarchy of the diagram.

The following subroutine shows the essential features of the Nim game:

GAME (SEQUENCE OF MOVES)

```
Display: Game position
REPEAT
        Player's move
        last move:= "human player"
        Display: Game position
        IF number of pieces of fruit = 0 THEN BREAK
        Computer's move
        last move:= "computer"
        Display: game position
        IF number of pieces of fruit = 0 THEN BREAK
END REPEAT
```

This tells us that the round consists of an exchange between a human player and the computer. In the variable "last move" the program remembers who moved last. This is important for determining the winner. The one to make the last move before the BREAK is the winner:

WINNING DECISION

```
IF last move = "human" THEN
        Display: "You win."
OTHERWISE
        Display: "I win."
END IF
```

This strategy of Bouton is expressed in the part of the program called "Computer's move": when the computer is about to move and is in a losing position, it makes a random move by taking one piece of fruit from a randomly chosen bowl.

176

```
Analysis of position
IF sum of columns = 0 THEN
        Random move
OTHERWISE
        Winning move
END IF
Display: "I'm taking .... pieces of fruit."
Count the fruit remaining
```

This has to be preceded by an analysis of the position: the number of pieces of fruit must be translated into binary numbers and then the columns of binary numbers must be added (according to the formula a (+) b = a+b - 2*a*b). The winning move can be described as follows: begin at the left, change the columns with an odd number of 1's by adding or removing a 1 so that an even number results. This is always possible by merely changing the amount in a single bowl. Accordingly, this process can be divided into the parts: "Seeking leading 1"; "Choosing the Bowl"; "Changing the Binary Numbers"; "Creating the Number to be Removed".

Bowl i	Position j (2^j)						Total pieces of fruit F(i)
	5 (32	4 16	3 8	2 4	1 2	0 1)	
1	0	0	0	0	1	1	3
2	0	0	0	1	0	1	5
3	0	0	1	1	0	1	13
Sum S(j)	0	0	1	0	1	1	

In this example 7 pieces of fruit are to be taken from the third bowl. Here is the program for the game:

```
01000 : PRINT CHR$(147)
01010 : PRINT "         THE GAME OF NIM
01020 : PRINT "         ---------------
01030 :
01040 REM THE COMPUTER PLAYS AGAINST THE USER WITH THE
01050 REM BOUTON STRATEGY OF NIM WITH 3 BOWLS
01060 :
```

```
01070 :
01080 REM *** MAIN PROGRAM *******************************
01090 :
01100 : GOSUB 1250 : REM RULES OF PLAY
01110 :
01120 : GOSUB 1400 : REM INITIALIZATION
01130 :
01140 : GOSUB 1550 : REM ROUND (SERIES OF MOVES)
01150 :
01160 : GOSUB 2920 : REM DETERMINING THE WINNER
01170 :
01180 : GOSUB 3000 : REM EXIT OR REPEAT
01190 :
01200 : END
01210 :
01220 REM *** END OF MAIN PROGRAM **************************
01230 :
01240 :
01250 REM ### SUBROUTINE 'RULES OF PLAY' ##################
01260 :
01270 : PRINT
01280 : PRINT "DO YOU WANT TO SEE THE RULES? (Y/N)
01290 : GET A$ : IF A$ = "" THEN 1290
01300 : IF A$ <> "Y" THEN RETURN
01310 :
01320 : PRINT
01330 : PRINT "ON THE TABLE ARE 3 BOWLS HOLDING PIECES OF
01340 : PRINT "FRUIT. YOU AND I WILL ALTERNATE TAKING AT
01350 : PRINT "LEAST ONE PIECE FROM ONE BOWL EACH TIME. THE
01360 : PRINT "LOSER IS THE ONE WHO CANNOT TAKE ANY MORE
01365 : PRINT "BECAUSE THE BOWLS ARE ALL EMPTY.
01370 :
01380 : RETURN : REM END OF GAME RULES ####################
01390 :
01400 REM ### SUBROUTINE 'INITIALIZATION' ################
01410 :
01420 : PRINT : PRINT
01430 : PRINT " THE AMOUNTS:
01440 : PRINT
01450 : LET N = 0       : REM TOTAL NUMBER OF PIECES OF FRUIT
01460 : FOR I = 1 TO 3
01470 :    PRINT "HOW MANY PIECES OF FRUIT ARE IN BOWL ";I;
01480 :    INPUT F
01490 :    IF F < 1 OR F > 50 THEN 1470
01500 :    LET F(I) = F : LET N = N+F
01510 : NEXT I
01520 :
01530 : RETURN : REM END OF INITIALIZATION ###############
01540 :
01550 REM ### SUBROUTINE 'ROUND' #########################
01560 :
01570 : GOSUB 1750                : REM DISPLAY OF GAME STATUS
```

```
01580 :
01590 : GOSUB 1890                : REM PLAYER'S MOVE
01600 :
01610 : GOSUB 1750                : REM DISPLAY OF GAME STATUS
01620 :
01630 : IF N <= 0 THEN RETURN : REM ALL BOWLS ARE EMPTY -
01635 :                            REM ROUND OVER
01640 :
01650 : GOSUB 2070                : REM COMPUTER'S MOVE
01660 :
01670 : GOSUB 1750                : REM DISPLAY OF GAME STATUS
01680 :
01690 : IF N <= 0 THEN RETURN : REM ALL BOWLS ARE EMPTY -
01695 :                            REM ROUND OVER
01700 :
01710 : GOTO 1590                 : REM JUMP BACK TO PLAYER'S MOVE
01720 :
01730 : REM END OF ROUND #################################
01740 :
01750 : REM +++ SUBROUTINE 'DISPLAY OF GAME STATUS' ++++++++
01760 :
01770 :    PRINT
01780 :    PRINT "CURRENT GAME STATUS:
01790 :    PRINT
01800 :    FOR I = 1 TO 3
01810 :       PRINT "BOWL ";I;": ";
01820 :       IF F(I) = 0 THEN 1840
01830 :       FOR J = 1 TO F(I) : PRINT "*      "; : NEXT J
01840 :       PRINT
01850 :    NEXT I
01860 :
01870 :    RETURN : REM END DISPLAY OF GAME STATUS ++++++++++
01880 :
01890 : REM +++ SUBROUTINE 'PLAYER'S MOVE' +++++++++++++++++
01900 :
01910 :    PRINT
01920 :    PRINT "YOUR TURN ";
01930 :    INPUT "WHAT BOWL ARE YOU TAKING FRUIT FROM? "; A
01940 :    IF A < 1 OR INT(A) <> A OR A > 3 THEN 1930
01950 :    IF F(A)=0 THEN PRINT "BOWL IS EMPTY." : GOTO 1930
01960 :
01970 :    INPUT "HOW MANY PIECES OF FRUIT "; F
01980 :    IF F < 1 OR INT(F) <> F OR F > F(A) THEN 1970
01990 :    LET F(A) = F(A)-F
02000 :
02010 :    LET N = F(1)+F(2)+F(3)
02020 :
02030 :    LET L$ = "PLAYER"
02040 :
02050 :    RETURN : REM END OF PLAYER'S MOVE ++++++++++++++
02060 :
02070 : REM +++ SUBROUTINE COMPUTER'S MOVE +++++++++++++++++
```

```
02080 :
02090 :    GOSUB 2210                        : REM ANALYZE STATUS
02100 :    IF S=0 THEN GOSUB 2520 : GOTO 2120: REM RANDOM
02105 :                                       REM MOVE
02110 :    GOSUB 2610
02120 :    PRINT
02130 :    PRINT "I AM TAKING ";Z;" PIECE(S) FROM BOWL ";I0
02140 :
02150 :    LET N = F(1)+F(2)+F(3)
02160 :
02170 :    LET L$ = "COMPUTER"
02180 :
02190 :    RETURN : REM END OF COMPUTER'S MOVE
02200 :
02210 :    REM +++ SUBROUTINE ANALYSIS OF GAME STATUS +++
02220 :
02230 :       REM POWERS OF TWO
02240 :
02250 :          LET P(0) = 1
02260 :          FOR J = 1 TO 5 : LET P(J) = 2*P(J-1) : NEXT J
02270 :
02280 :       REM BINARY REPRESENTATION
02290 :
02300 :          FOR I = 1 TO 3
02310 :            LET M = F(I)
02320 :            FOR J = 5 TO 0 STEP -1
02330 :              LET P = P(J) : LET Z = INT(M/P)
02340 :              LET D(I,J) = Z : LET M = M - Z*P
02350 :            NEXT J
02360 :          NEXT I
02370 :
02380 :       REM FORM COLUMN TOTALS
02390 :
02400 :          LET S = 0
02410 :          FOR J = 0 TO 5
02420 :            LET S(J) = 0
02430 :            FOR I = 1 TO 3
02440 :              LET D = D(I,J)
02450 :              LET S(J) = S(J)+D - 2*S(J)*D
02460 :            NEXT I
02470 :            LET S = S + S(J)
02480 :          NEXT J
02490 :
02500 :          RETURN : REM END OF STATUS ANALYSIS
02510 :
02520 :    REM +++ SUBROUTINE RANDOM MOVE +++
02530 :
02540 :       LET I0 = INT(3*RND(TI))+1  : REM CHOICE OF BOWL
02550 :       IF F(I0) = 0 THEN 2540    : REM BOWL EMPTY
02560 :       LET Z = 1                 : REM AMOUNT REMOVED
02570 :       LET F(I0) = F(I0) - Z     : REM NEW FRUIT TOTAL
02580 :
```

```
02590 :     RETURN : REM END OF RANDOM MOVE
02600 :
02610 :   REM +++ SUBROUTINE WINNING MOVE +++
02620 :
02630 :     REM LOOK FOR LEADING 1
02640 :
02650 :       FOR J = 5 TO 0 STEP -1
02660 :         IF S(J) > 0 THEN LET J0 = J : LET J = 0
02670 :       NEXT J
02680 :
02690 :     REM BOWL CHOICE
02700 :
02710 :       FOR I = 1 TO 3
02720 :         IF D(I,J0) > 0 THEN LET I0 = I : LET I = 3
02730 :       NEXT I
02740 :
02750 :     REM CHANGE BINARY REPRESENTATION
02760 :
02770 :       LET D(I0,J0) = 0
02780 :       IF J0 = 0 THEN 2830 : REM CHANGE 'FIRST'
02785 :                             REM COLUMN ONLY
02790 :       FOR J = J0-1 TO 0 STEP -1
02800 :         IF S(J) > 0 THEN LET D(I0,J) = 1 - D(I0,J)
02810 :       NEXT J
02820 :
02830 :     REM GENERATE AMOUNT TO BE REMOVED
02840 :
02850 :       LET K = 0
02860 :       FOR J = 0 TO 5 : LET K = K+P(J)*D(I0,J):NEXT J
02870 :       LET Z = F(I0)-K     : REM AMOUNT REMOVED
02880 :       LET F(I0) = K
02890 :
02900 :     RETURN : REM END OF WINNING MOVE
02910 :
02920 REM ### SUBROUTINE 'DETERMINING THE WINNER' #########
02930 :
02940 : PRINT
02950 : IF L$ = "COMPUTER" THEN PRINT "I WIN." : RETURN
02960 :                            PRINT "YOU WIN."
02970 :
02980 : RETURN : REM END DETERMINING THE WINNER ###########
02990 :
03000 REM ### SUBROUTINE 'EXIT OR REPEAT' ###############
03010 :
03020 : PRINT : PRINT
03030 : PRINT "ANOTHER ROUND? (Y/N)
03040 : GET A$ : IF A$ = "" THEN 3040
03050 : IF A$ = "Y" THEN RUN
03060 :
03070 : PRINT : PRINT : PRINT
03080 : PRINT "GOOD-BYE."
03100 : RETURN : REM END OF EXIT OR REPEAT ###############
```

Example 6.2 BOWLING

A line of bowling pins is divided into groups:

You can roll a ball at them and knock down a single pin or
two adjacent pins (which removes them from play). Pins in
two different groups are not considered adjacent. The loser
is the player who can't move (in this case, bowl) any more
because all the pins are gone. The object is to devise a
winning strategy.

Let's assume that at the beginning of the game five pins are
standing in a group (now graphically represented by small
circles: ooooo). What can player A, who now has to move,
do? He can remove one or two pins, either from inside the
group or else from the outer edges. We can express the
possibilities like this:

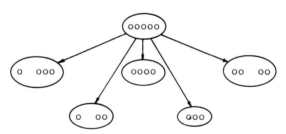

Only one of these five possibilities will be winning (if the
opponent doesn't make a mistake), that move being to position (oo oo). Whatever player B now tries, A can copy him
and will thereby get the last move. Let's say (as we did
with the game of Nim) that (oo oo) is a losing position for
the player just taking a turn. In order to show that the
other four positions are winning positions, we can make the
diagram more complete. This diagram is called a graph; the
circles are called "knots" and the arrows -----> "directed
edges".

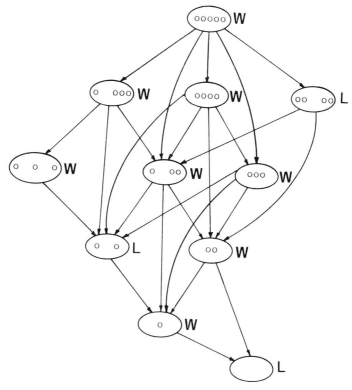

This diagram tells us the following:

1. The final knot is an L-knot, for a losing position.
2. From each W-knot at least one arrow leads to an L-knot.
3. From each L-knot any arrow leads to a W-knot.

We have to imagine a round in the game proceeding as if a player were moving a piece along a path from knot to knot along the directed edges in the direction of the arrows. The loser is then the player who cannot move any longer because his piece has arrived at the final knot. The player who begins in a W-knot possesses a winning strategy! All he has to do is move to an L-knot (which always works according to number 2, above). At the next move the other player has to move to a W-knot (according to number 3).

Is it possible to tell from the playing positions whether they are winning or losing positions as we could in the Nim game? Let's take an example. Of course the losing

183

positions (o o) and (oo oo) could be construed as "symetrically balanced"; this balance is apparent from the Nim totals:

1	10
<u>1</u>	<u>10</u>
0	00

But (o oooo) is also a losing position. Notice that all the arrows leading from this knot lead to winning positions. The following graph shows this. Our new system of abbreviation writes (o oooo) as (1,4).

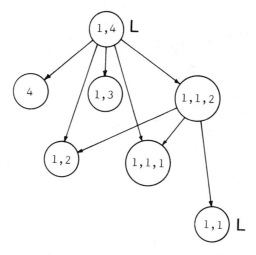

If we want to preserve our criterion of symmetry ("Nim total zero"), then we mustn't take the binary representation of the number of bowling pins, but rather find other numbers to represent the positions.

The Berlin mathematician Roland Sprague got the breakthrough idea to solve this in 1936 (as did P. M. Grundy independently of him in 1939). He assigned to each position (every knot on the game diagram), an integer called its "rank."

This is how it works:

1. The end knot (or knots) receives the rank of zero (0).

2. The rank of a knot is the smallest integer not assigned to the immediately following knots.

(Knot Q immediately follows P if an arrow leads from P to Q.) It is important to note that the rank of a knot is different from the rank of all those immediately following.

With the help of the Sprague rules 1 and 2 we can redefine our previous game diagram assigning ranks (written to the right of each knot):

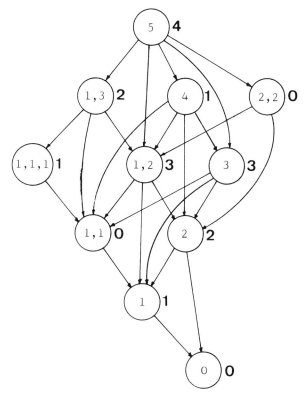

You can tell from this that a knot is a losing knot if it has the rank of 0. The reasons are:

1. End knots have the rank of 0.

2. There is always an arrow from a given knot with rank > 0 to a knot with rank 0. If there weren't such an arrow, the given knot (according to Sprague's rule 2) would have to get the rank of 0 itself.

3. Every arrow from a knot ranked 0 leads to a knot with rank > 0 because the rank of a knot is different from every one of its immediate successors on the sequence.

185

We can tell a few things about our game: rank(1) = 1, rank(2) = 2, rank(3) = 3, rank(4) = 1, rank(5) = 4, rank(6) = 3.

Something else is also apparent from the game diagram: rank(1,3) = 2; rank(1) = 1; rank(3) = 3; and 1 (+) 3 = 2. Therefore:

rank(1,3) = rank(1) (+) rank(3).

Furthermore:

rank(2,2) = 0; rank(2) = 2; and 2 (+) 2 = 0;

Therefore:

rank(2,2) = rank(2) (+) rank(2).

This is a relationship that holds true in general (though we can't go into the proof): <u>the rank of a game move is equal to the Nim total of the ranks of its components. A losing position is said to have the ranks of its components equal to 0.</u> In order to identify the losing positions we need tables of ranks like these:

n	0	1	2	3	4	5	6	7	8	9
rank(n)	0	1	2	3	1	4	3	2	1	4

n	10	11	12	13	14	15	16	17
rank(n)	2	6	4	1	2	7	1	4

The ranks are also called <u>Sprague-Grundy numbers</u> and the function that assigns a Sprague-Grundy number to a position is called the <u>Sprague-Grundy Function.</u>

We now know that (to pick one example) for the position (2,4,5) there is a position with the rank of 0 that can be reached in one move. We have to present this position to our opponent if we want to win. But we don't know how to find it yet! This time it's not as easy as the Nim game was. If you are interested in pursuing this problem, generate all the knots that immediately follow a position and examine them for a position with the rank of 0. The theory tells us that this process will always work. The program is left up to you.

****

Example 6.3   QUEEN IN THE CORNER

Player A puts the queen in one of the fields in the upper-most row or in the column farthest right (shaded in the dia-gram).

The queen can move as usual but only in the directions west, southwest, and south.   Player B moves first and then the players alternate.   The one who can no longer move because the queen is in the corner (marked in diagram with *), is the loser.   The object is to find a winning strategy.

We proceed from the end position (0,0) and construct the Sprague-Grundy Function and a winning-losing breakdown. First we determine that all positions (0,1), (0,2), (0,3), ... , (1,0), (2,0), (3,0), ... as well as the diagonal (1,1), (2,2), (3,3), etc. are winning positions (for the player about to take his turn).   Their Sprague-Grundy num-bers are:

5	–				
4	–				
3	–				
2	[0]	–			
1	2	[0]	–	–	–
0	1	2	3	4	5

The two first losing positions are (2,1) and (1,2).   Their Sprague-Grundy numbers are 0 (see above).   If we proceed recursively (assigning one position up the diagonal and then

187

to either side) then we get the following table (losing positions are marked):

7	8	6	9	[0]	1	4	5
6	7	8	1	9	10	3	4
5	3	4	[0]	6	8	10	1
4	5	3	2	7	6	9	[0]
3	4	5	6	2	[0]	1	9
2	[0]	1	5	3	4	8	6
1	2	[0]	4	5	3	7	8
0	1	2	3	4	5	6	7

If we had continued the table we could have discovered the following losing positions:

n	1	2	3	4	5	6	7	8	9	10	11	12	13	14
x(n)	1	3	4	6	8	9	11	12	14	16	17	19	21	22
y(n)	2	5	7	10	13	15	18	20	23	26	28	31	34	36

Comment: We have used only the coordinates of the losing positions right of the diagonal; the others are located symmetrically to it, which means that x(n) and y(n) are reversed.

Does any mathematical rule apply here? A couple of things obviously are true:

$$(1) \quad y(n) = x(n) + n$$
$$(2) \quad x(y(n)) = x(n) + y(n).$$

Knowing this we can easily calculate the losing positions numerically:

$$x(1) = 1$$
$$x(n) = \text{smallest integer not among}$$
$$x(k), y(k) \text{ when } k < \dot{n}$$
$$y(n) = x(n) + n$$

Lots of exciting mathematical discoveries could be made from the results of x(n) and y(n). For example, you can see members of the well-known Fibonacci series. Space unfortunately does not permit us to explore this here, but the activities that follow may inspire you.

BRAINTEASERS

1. Expand the program for the Nim game to include any number of bowls.

2. Alter the Nim game so that: 1) the beginning number of pieces of fruit in a bowl is generated at random; and 2) the player to move next (in a game with more than 2 players) is determined at random.

3. Show that the Nim total of the number of pieces of fruit in the Nim game is in each case the Sprague-Grundy number of each game position.

4. Look at the triangle pictured here. Show that for every triple whose numbers are on a straight line, each number is the Nim total of the other two. Example: 2 (+) 5 = 7; 2 (+) 7 = 5; 5 (+) 7 = 2. Knowing this is helpful in the Nim game. Do you know why?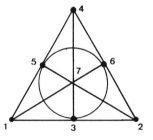

5. Show that the set of natural numbers forms a group with the operation of Nim addition. Why not construct a table of relationships? The beginning looks like this:

(+)	0	1	2	3	4	5	6	7	8
0	0	1	2	3	4	5	6	7	8
1	1	0	3	2	5	4	7	6	9
2	2	3	0	1	6	7	4	5	10
3	3	2	1	0	7	6	5	4	11
4	4	5	6	7	0	1	2	3	12
5	. . . .

6. NECKLACE

On the table are N coins placed in a circle.

They are arranged so that each touches its neighbor. Two
players take turns removing one or two adjacent coins from
the circle. The player who can no longer move because all
coins are gone is the loser. Prove that a "symmetry strategy" underlies this game and that it is constructed in the
following way: With the second move the player whose turn it
is changes the open chain into two partial chains of equal
length. Then he continues to do the same thing to one chain
that his opponent does to the other. Who ultimately wins
with such a strategy? Write a program for the game.

7. "NO REPEAT NIM"

This game has the same rules of play as Nim (played with
one bowl) yet has one limitation. The previous moves of the
opponent may not be repeated. Use the Sprague-Grundy Function to develop a winning strategy.

8. Write a program for the game of BOWLING.

9. Write a program for the game QUEEN IN THE CORNER.

10. TAKING STONES

There is an old Chinese game called "Picking up Pebbles"
(Jian-shi-zi). It goes like this: there are two piles of
pebbles. Players can take pebbles alternately from one pile
or the other or they may take from both piles at once --

190

just as long as it's the same number of stones from each pile.

Demonstrate that this game is essentially the same game as "Queen in the Corner." This game was reinvented in 1907 by the Dutchman W.A. Wythoff and has been called "Wythoff's Nim" after him.

11. There is a pile of N objects (matches, pebbles, jelly beans). The player whose move it is takes away as many as he likes (but not the whole pile). Then the players alternate taking one or more objects, but each time removing twice the number that the other player removed in his previous move. Use the Sprague-Grundy Function to develop a winning strategy for this game.

12. LASKER'S NIM
The world chess champion Emanuel Lasker wrote a book in 1931 called Board Games of Different Peoples in which he describes a variation on the Nim game: "The player whose move it is may divide any pile into two, or reduce one pile in size by any amount." The winner is the player who moves last. Lasker calculated a few losing positions but then wrote: "I have not found the law of losing positions but it would seem to be associated with the Nim game." You have the advantage of knowing the Sprague-Grundy Function and can calculate this "law of losing positions" as well as develop a strategy.

13. There is a row of squares any length you wish. On these separate squares are lying N stones. Two players take turns moving any stone to any free square up toward the beginning of the row. Jumping other stones is (a) allowed or (b) not allowed. The player who can no longer move a stone because all the squares at the beginning of the row are filled with stones is the loser.

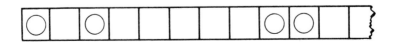

The game is called "Moving Stones." If you are playing with the (b) variation it's also called "Welter's Game." Your object is to show that it is a variation of the Nim game. Use the Sprague-Grundy Function and find a winning strategy that can be taught to the computer.

14. MONEY BAGS
This game goes like the Moving Stones except that it is played with coins and a silver dollar. The moves are just like Moving Stones. On the left is a money bag to catch the coins that fall off the row of squares when one is moved to the far left. The winner is the player who can move the silver dollar into the money bag.

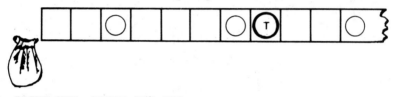

15. PRIME NIM, PRIME AND DIM
Prime Nim is Nim with the added restriction that the piles can only be changed by prime numbers. In other words the number of objects (fruit, stones, matches) removed must be a prime number. 1 is counted here as a prime number.

In the game called Prime the numbers of objects to be removed must be relatively prime (have no common divisor) to the number in the pile. Example: if 8 objects are lying in a pile then only 1, 3, 5, or 7 of them can be removed.

In the game Dim, on the other hand, the number of objects to be removed must be a divisor of the number in the pile.

Figure out the Sprague-Grundy Functions as well as the winning strategy for each game and write the programs to accompany them.

16. NIM WITH A CATCH
E.H. Moore devised this catch to the Nim game: the number of objects to be removed cannot exceed the number K. Show that a position is a losing position if its Nim total is 0

when divided by K+1. Build upon this to construct a winning strategy and to write a program.

17. DISTICH (GRUNDY'S GAME)
As in Nim we have one (or more) pile of objects. A move consists of dividing the pile into two unequal piles. The loser is the player who cannot move because there is nothing more to divide. Generate the Sprague-Grundy Function and develop a winning strategy for the computer.

18. RINGS
The starting situation in this game by J.H. Conway consists of a few dots on paper. In order to move, simply draw a ring through some of these. The ring may not touch or cross over itself or another ring. You have to show that this is equivalent to the Nim game with the additional rule that by removing objects from inside a pile, new piles can be constructed. For example, this position in the Rings game:

corresponds to the position (3,5,2) in the Nim game. If each ring has to go through exactly one or two of the dots, then we have a version of the game that is equivalent to the Bowling game (6.2).

19. Two-dimensional Nim
Here is another Nim version by D.Gale. The playing stones are lying on a board with M rows and N columns. To move, select the position io, jo (there has to be a stone on these coordinates) and then remove all the stones with positions i >= io and j >= jo. The player who is forced to choose the position 1,1 loses.

20. TAC-TIX
The Dane Piet Hein invented the following variation on the

193

Nim game. The stones lie on on a board M x N just as in
Two-dimensional Nim.   Players take turns removing stones
from a horizontal row or a vertical column with the only
added catch that the stones have to be in adjacent squares.

In a typical game, the loser is the player who can't
remove any more stones. Show that a normal game, in which a
player loses when he can't remove any more stones, has a
simple winning strategy and is therefore fairly uninterest-
ing. Analyze the variation of the game using a 3x3 quadrant
and develop a winning strategy for this one.

****

One of the oldest games known is Tic Tac Toe. It has
even been found on Egyptian wall paintings and Greek vases.
King Tutankamen, Socrates and Queen Elizabeth I would all
have recognized the figures on the postage stamp above and
known how to play the game. Charles Babbage, one of the
fathers of the computer, developed six different machines to
play Tic Tac Toe.

There are countless strategies and computer programs
for the game. You may think the game is easy to play but it
poses a programming problem of no small dimension.

194

Example 6.4   TIC TAC TOE

A player "Tic" (O) and a player "Tac" (X) take turns filling the squares of a 3x3 quadrant with the object of com-
pleting any straight line (three in a row) with one's own
symbol.   This can be vertical, horizontal, or diagonal.
You'll be playing against the computer.

This is how a possible round might look:

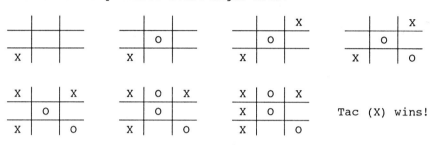

Tac (X) wins!

On the next page is the rough structure of a program that
looks like one for any game.   But the section "move selection" is the interesting feature because it contains the
strategy used by the computer.   The object is to give every
playing position a computer move.   Here are the goals of the
program according to priority.

(1) Generate a win by filling a series (row, column, or
diagonal) with three of the player's own characters.
(2) Prevent an opponent's win.
(3) Fill a row with two of one's own characters.
(4) Prevent the opponent from doing this.

   The first two partial goals are easy to program.   We
have to test whether there is a row with a "hole" (i.e., two
identical pieces and an empty field) and then fill this hole
with one's own symbol (for the computer a O).   The other
goals go into a part of the program called the evaluation
function.   This is a directive that gives every playing

position a number (actually its "value" in relation to the goals.)

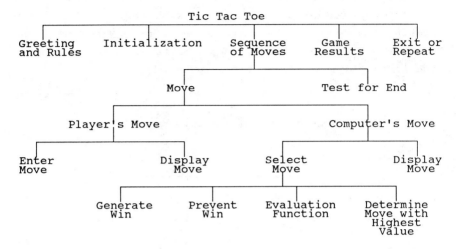

The computer takes the move that has the highest evaluative function value. How does the computer know whether it is threatened by a win? Very simple: we create a table T in which the eight possible wins are stored and another table U containing playing positions. When you play with the program, alter table U and study how the computer's evaluation changes!

Table B contains the playing field with its nine quadrants.

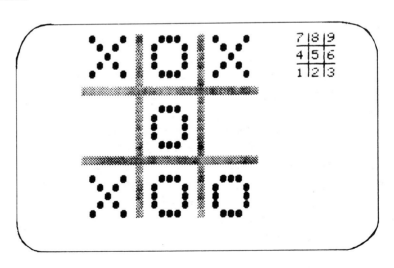

To which the following table corresponds:

1	2	3	4	5	6	7	8	9
-1	1	1	0	1	0	-1	1	-1

After the player's or the computer's move either 1 or -1 is
entered into the table. The "flag" F determines the value:
it is -1 when a player moves, and 1 for the computer's move.

The program is schematic enough (I hope) so that you
can understand it easily. You'll be able to figure out the
evaluation functions after a few tries.

```
01000 : PRINT CHR$(147)
01010 : PRINT "          TIC-TAC-TOE
01020 : PRINT "          -----------
01030 :
01040 REM THE COMPUTER WILL PLAY TIC-TAC-TOE (USING O'S)
01050 :
01060 :
01070 REM *** MAIN PROGRAM *********************************
01080 :
01090 : GOSUB 1270 : REM GREETING AND GAME DESCRIPTION
01100 :
01110 : GOSUB 2000 : REM INITIALIZATION
01120 :
01130 : GOSUB 2600 : REM DISPLAY OF PLAYING AREA
01140 :
01150 : GOSUB 2870 : REM GAME
01160 :
01170 : GOSUB 4330 : REM DETERMINING THE WINNER
01180 :
01190 : GOSUB 4450 : REM EXIT OR REPEAT
01200 :
01210 : END
01220 :
01230 REM *** END OF MAIN PROGRAM *************************
01240 :
01250 :
01260 :
01270 REM ### PROCEDURE 'GREETING AND GAME DESCRIPTION' ####
01280 :
01290 : PRINT : PRINT
01300 : INPUT "WHAT IS YOUR FIRST NAME "; N$
01310 : PRINT
01320 : PRINT "HELLO "; N$;" !" : PRINT
01330 : PRINT "WOULD YOU LIKE TO READ THE RULES? (Y/N)
01340 : GET A$ : IF A$ = "" THEN 1340
```

```
01345 : IF A$ <> "Y" THEN RETURN
01350 :
01360 : REM INSERT RULES OF PLAY HERE
01370 :
01950 : RETURN
01960 :
01970 REM ### END OF GREETING AND GAME DESCRIPTION #########
01980 :
01990 :
02000 REM ### PROCEDURE 'INITIALIZATION' #################
02010 :
02020 : REM --- SETTING UP THE TABLES
02030 :
02040 : DIM T(8,3)    : REM TABLE OF WINNING CONFIGURATIONS
02050 : DIM U(9,4)    : REM TABLE OF SPECIAL GAME POSITIONS
02060 : DIM C(9)      : REM COUNTER FOR IDENTICAL CHARACTERS
02070 : DIM B(9)      : REM TABLE SHOWING OCCUPIED SPACES
02080 : DIM V(9)      : REM TABLE FOR EVALUATION
02090 : DIM C$(2)     : REM CROSSES AND CIRCLES
02100 : DIM L(9)      : REM COORDINATES OF THE PLAYING SPACES
02110 :
02120 : REM --- ENTERING THE PLAYING POSITIONS
02130 :
02140 :    FOR I = 1 TO 8
02150 :       FOR J = 1 TO 3 : READ T(I,J) : NEXT J
02160 :    NEXT I
02170 :
02180 DATA 7,8,9,4,5,6,1,2,3  : REM HORIZONTAL SERIES
02190 DATA 7,4,1,8,5,2,9,6,3  : REM VERTICAL SERIES
02200 DATA 7,5,3,1,5,9        : REM DIAGONAL SERIES
02210 :
02220 :    FOR I = 1 TO 9
02230 :       FOR J = 1 TO 4 : READ U(I,J) : NEXT J
02240 :    NEXT I
02250 :
02260 DATA 3,4,8,0,3,5,0,0,3,6,7,0
02270 DATA 2,4,0,0,2,5,7,8,2,6,0,0
02280 DATA 1,4,7,0,1,5,0,0,1,6,8,0
02290 :
02300 : REM --- X AND O
02310 :
02320 :    LET T$ = "O"         : LET T$ = T$ + CHR$(17)
02321 :    LET T$ = T$ + T$     : LET T$ = T$ + T$
02322 :    LET T$ = T$ + "O"    : LET T$ = T$ + CHR$(157)
02323 :    LET A$ = CHR$(157)   : LET A$ = A$ + A$
02324 :    LET A$ = A$ + A$     : LET A$ = T$ + A$
02325 :    LET A$ = "O"         : LET A$ = A$ + CHR$(145)
02326 :    LET T$ = T$ + A$     : LET A$ = A$ + A$
02327 :    LET A$ = A$ + A$     : LET T$ = T$ + A$
02328 :    LET C$(1) = T$
02329 :
02330 :    LET T$ = CHR$(29) : LET T$ = T$ + T$
```
                                    198

```
02331 :    LET A$ = "○" + CHR$(157) + CHR$(157)
02332 :    LET T$ = T$ + A$ + A$ + " "
02333 :    LET A$ = CHR$(17) + CHR$(157) + "○"
02334 :    LET T$ = T$ + A$ + A$ + A$
02335 :    LET T$ = T$ + CHR$(17) + "○○○"
02336 :    LET A$ = CHR$(145) + "○" + CHR$(157)
02337 :    LET T$ = T$ + A$ + A$ + A$
02338 :    LET T$ = T$ + CHR$(157) + CHR$(145) + "○ "
02339 :    LET C$(2) = T$
02340 :
02350 : REM --- COORDINATES OF THE 9 FIELDS
02360 :
02370 :    FOR I = 1 TO 9 : READ L(I) : NEXT I
02380 :
02390 DATA 1608,1616,1624,808,816,824,8,16,24
02400 :
02410 : REM --- GRID
02420 :
02430 :    LET G1$ = "                       ░                ░
02440 :    LET G2$ = "          ▓▓▓▓▓▓▓▓▓▓▓▓▓▓▓▓▓▓▓▓▓▓▓▓▓▓▓▓▓▓▓▓
02450 :
02460 : REM --- CLEAR PLAYING GRID
02470 :
02480 :    FOR I = 1 TO 9 : LET B(I) = 0 : LET C(I) = 0
02485 :    NEXT I
02490 :
02500 :
02510 : REM --- CHARACTER SEQUENCE FOR CURSOR COMMANDS
02520 :
02530 :    LET D$ = ""
02532 :    FOR T = 1 TO 23
02534 :      LET D$ = D$ + CHR$(17)
02536 :    NEXT T
02540 :
02550 : RETURN
02560 :
02570 REM ### END OF INITIALIZATION #######################
02580 :
02590 :
02600 REM ### PROCEDURE DISPLAY PLAY AREA #################
02610 :
02620 : REM --- GRID
02630 :
02640 :    PRINT CHR$(147);
02650 :
02660 :    FOR J = 1 TO 6 : PRINT G1$ : NEXT J
02670 :    PRINT G2$
02680 :    FOR J = 1 TO 7 : PRINT G1$ : NEXT J
02690 :    PRINT G2$
02700 :    FOR J = 1 TO 6 : PRINT G1$ : NEXT J
02710 :    PRINT G1$
02720 :
```

```
02730 :    PRINT CHR$(19);
02740 :
02750 : REM --- SMALL PLAYING AREA UPPER RIGHT
02760 :
02770 :    FOR J = 1 TO 34 : PRINT CHR$(29); : NEXT J
02771 :    LET A$ = CHR$(157) : LET A$ = A$ + A$
02772 :    LET A$ = A$ + A$ + CHR$(157)
02773 :    LET T$ = CHR$(17) + A$
02774 :    LET A$ = "7|8|9" + T$ + "-|-|-" + T$ + "4|5|6"
02775 :    LET A$ = A$ + T$ + "-|-|-" + T$ + "1|2|3"
02780 :    PRINT A$
02790 :
02800 :    PRINT CHR$(19); D$;
02810 :
02820 : RETURN
02830 :
02840 REM ### END OF PLAY AREA DISPLAY ####################
02850 :
02860 :
02870 REM ### PROCEDURE 'GAME'  #########################
02880 :
02890 : REM = SHORT PROCEDURE 'SEQUENCE OF MOVES' ==========
02900 :
02910 :    FOR N = 1 TO 5
02920 :
02930 :       GOSUB 3110              : REM PLAYER'S MOVE
02940 :
02950 :       GOSUB 4060              : REM TEST FOR END
02960 :
02970 :       IF E$ = "DONE" THEN RETURN   : REM END OF ROUND
02980 :
02990 :       GOSUB 3370              : REM COMPUTER'S MOVE
03000 :
03010 :       GOSUB 4060              : REM TEST FOR END
03020 :
03030 :       IF E$ = "DONE" THEN RETURN   : REM END OF ROUND
03040 :
03050 :    NEXT N
03060 :
03070 :    RETURN
03080 :
03090 : REM = END OF SHORT PROCEDURE 'SEQUENCE OF MOVES'====
03100 :
03110 : REM +++ SUBPROCEDURE 'PLAYER'S MOVE' +++++++++++++++
03120 :
03130 :    REM - SHORT PROCEDURE 'PLAYER'S MOVE' ------------
03140 :
03150 :       LET F = -1 : REM FLAG FOR PLAYER
03160 :       GOSUB 3220 : REM ENTER MOVE
03170 :       GOSUB 3930 : REM ENTER INTO GRID
03180 :       RETURN
03190 :
```

```
03200 :    REM - END OF SHORT PROCEDURE 'PLAYER'S MOVE' -----
03210 :
03220 :    REM +++ SUBPROCEDURE 'ENTER MOVE' ++++++++++++++++
03230 :
03240 :      PRINT CHR$(19); D$; " YOUR MOVE? (DIGIT 1 TO 9)
03250 :      GET Z$ : IF Z$ = "" THEN 3250
03260 :
03270 :      IF Z$ < "1" OR Z$ > "9" THEN 3240
03280 :
03290 :      LET Z = VAL(Z$)
03300 :      IF B(Z) <> 0 THEN 3240 : REM FIELD ALREADY
03305 :                               REM OCCUPIED
03310 :      RETURN
03320 :
03330 :    REM +++ END OF 'ENTER MOVE' +++++++++++++++++++++++
03340 :
03350 : RETURN : REM +++ END OF PLAYER'S MOVE +++++++++++++
03360 :
03370 : REM +++ SUBPROCEDURE 'COMPUTER'S MOVE' +++++++++++++
03380 :
03390 :    REM - SHORT PROCEDURE 'COMPUTER'S MOVE' ----------
03400 :
03410 :      LET F = 1   : REM FLAG FOR COMPUTER
03420 :      GOSUB 3480 : REM CHOOSE MOVE
03430 :      GOSUB 3930 : REM ENTER INTO GRID
03440 :      RETURN
03450 :
03460 :    REM - END OF SHORT PROCEDURE 'COMPUTER'S MOVE'----
03470 :
03480 :    REM +++ SUBPROCEDURE 'CHOOSE MOVE' ++++++++++++++
03490 :
03500 :      REM --- GENERATE WIN
03510 :
03520 :        FOR I = 1 TO 8
03530 :          IF C(I) = 2 THEN LET I1 = I : GOTO 3640
03540 :        NEXT I
03550 :
03560 :      REM --- PREVENT WIN
03570 :
03580 :        FOR I = 1 TO 8
03590 :          IF C(I) = -2 THEN LET I1 = I : GOTO 3640
03600 :        NEXT I
03610 :
03620 :        GOTO 3710 : REM TO EVALUATIVE FUNCTION
03630 :
03640 :      REM --- FILL HOLE
03650 :
03660 :        FOR J = 1 TO 3
03670 :          LET Z = T(I1,J)
03680 :          IF B(Z) = 0 THEN RETURN : REM HOLE FOUND
03690 :        NEXT J
03700 :
```

```
03710 :     REM --- EVALUATIVE FUNCTION
03720 :
03730 :       FOR I = 1 TO 9
03740 :         LET V(I) = 0
03750 :         IF B(I) <> 0 THEN 3800
03760 :         FOR J = 1 TO 4
03770 :           LET P = U(I,J)
03775 :           LET TT = ABS(C(P)) + 1
03780 :           IF P > 0 THEN LET V(I) = V(I) + TT
03790 :         NEXT J
03800 :       NEXT I
03810 :
03820 :     REM --- DETERMINE MOVE OF HIGHEST VALUE
03830 :
03840 :       LET V = 0
03850 :       FOR I = 1 TO 9
03860 :         IF V(I) <= V THEN 3890
03870 :         LET V = V(I) : REM HIGHEST VALUE
03880 :         LET Z = I    : REM MOVE WITH HIGHEST VALUE
03890 :       NEXT I
03900 :
03910 : RETURN : REM END OF CHOOSING MOVE +++++++++++++++++++
03920 :
03930 : REM +++ SUBPROCEDURE 'ENTERING INTO GRID' ++++++++++
03940 :
03950 :    LET B(Z) = F : REM FILLING FIELD NO. Z
03955 :                   REM WITH 1 OR -1
03960 :
03970 :    PRINT CHR$(19);
03980 :    LET I = INT(L(Z)/100) : REM ROW
03990 :    LET J = L(Z) - 100*I  : REM COLUMN
04000 :    FOR I1 = 1 TO I
04003 :       IF I > 0 THEN PRINT CHR$(17);
04006 :    NEXT I1
04010 :    FOR J1 = 1 TO J : PRINT CHR$(29); : NEXT J1
04020 :    PRINT C$((B(Z)+3)/2)  : REM X OR O
04030 :
04040 : RETURN : REM END OF FILLING THE GRID ++++++++++++++
04050 :
04060 : REM +++ SUBPROCEDURE 'TEST FOR END' ++++++++++++++++
04070 :
04080 :    REM --- HAS SERIES BEEN FOUND?
04090 :
04100 :       FOR J = 1 TO 4
04110 :         LET P = U(Z,J)
04120 :         IF P = 0 THEN 4160
04130 :         LET C(P) = C(P) + F
04140 :         LET C = C(P)
04150 :         IF C=3 OR C = -3 THEN LET E$="DONE" : RETURN
04160 :       NEXT J
04170 :
04180 :    REM --- ALL FIELDS OCCUPIED?
```

```
04190 :
04200 :       FOR I = 1 TO 9
04210 :       IF B(I) = 0 THEN RETURN : REM ANOTHER FREE FIELD
04215 :                                 REM AVAILABLE
04220 :       NEXT I
04230 :
04240 :       LET E$ = "DONE"
04250 :
04260 :       RETURN
04270 :
04280 : REM END OF TEST FOR END ++++++++++++++++++++++++++++
04290 :
04300 REM ### END OF PROCEDURE 'GAME' ####################
04310 :
04320 :
04330 REM ### PROCEDURE 'DETERMINE WINNER' ###############
04340 :
04350 : PRINT CHR$(19); D$;
04360 :
04370 : IF C = -3 THEN PRINT "*** YOU HAVE WON ***" : RETURN
04380 : IF C = 3 THEN PRINT "*** YOU HAVE LOST ***" : RETURN
04385 : PRINT CHR$(18)"      IT'S A DRAW
04390 :
04400 : RETURN
04410 :
04420 REM ### END OF 'DETERMINE WINNER' ##################
04430 :
04440 :
04450 REM ### EXIT OR REPEAT #############################
04460 :
04470 : FOR I = 1 TO 5000 : NEXT I : REM WAIT LOOP
04480 :
04490 : PRINT CHR$(147) : PRINT : PRINT : PRINT
04500 : PRINT "     ANOTHER ROUND? (Y/N)
04510 :
04520 : GET A$ : IF A$ = "" THEN 4520
04530 : IF A$ = "Y" THEN RUN
04540 :
04550 : PRINT : PRINT : PRINT : PRINT " I ENJOYED THE GAME.
04560 :
04570 : RETURN
04580 :
04590 REM ### END OF EXIT OR REPEAT
04600 :
04610 :
```

22. You may have discovered that it is possible to win the
game of Tic-Tac-Toe (6.4). Change the contents of U (lines
2260-2280) or the evaluative function (line 3780) or both so
that the computer never loses.

23. Expand the program for Tic-Tac-Toe so that the player
about to move is a) determined by chance, or b) can be determined by the user.

24. Write a program so two people can play Tic-Tac-Toe. The
computer sits out the action as Game Master.

25. The following Tic-Tac-Toe program is poorly structured.
This makes it barely understandable. Go ahead and try to
wade through it anyway and give it a clear structure. Hint:
reorganize it according to the phases of the game.

```
00100 PRINT CHR$(147)
00110 PRINT "          TIC-TAC-TOE
00120 PRINT "          -----------
00130 :
00140 REM A POORLY-STRUCTURED PROGRAM, MEANT
00145 REM AS A NEGATIVE EXAMPLE.
00150 :
00160 PRINT : PRINT
00170 PRINT "THE COMPUTER USES '0', YOU USE '11'.
00180 PRINT "EVERY FIELD WILL BE FILLED WITH A NUMBER FROM
00190 PRINT " 1-9. THE COMPUTER PLAYS FIRST.  YOUR MOVE
00200 PRINT " CONSISTS OF ENTERING THE NUMBER OF THE FIELD
00205 PRINT " YOU WISH TO OCCUPY.
00210 PRINT : PRINT
00220 :
00230 FOR I = 1 TO 9 : LET T(I) = I : NEXT I
00240 LET B = 9
00250 DEF FN A(X) = X - 8*INT((X-1)/8)
00260 GOSUB 500
00270 GOSUB 570
00280 LET J = 0
00290 LET J = J+1
```

```
00300 IF J = 4 THEN 380
00310 LET B = FNA(P+1)
00320 GOSUB 500
00330 GOSUB 570
00340 IF P = FNA(B+4) THEN 290
00350 LET B = FNA(B+4)
00360 GOSUB 500
00370 GOTO 400
00380 LET B = FNA(P+5)
00390 PRINT : PRINT "IT'S A DRAW." : GOTO 410
00400 PRINT : PRINT "THE COMPUTER WINS."
00410 PRINT : PRINT "ANOTHER ROUND? (Y/N)
00420 GET A$ : IF A$ = "" THEN 420
00430 IF A$ = "Y" THEN 230
00440 PRINT
00450 PRINT "            TOO BAD!
00460 END
00470 :
00500 REM SUBROUTINE 1
00510 LET T(B) = 0
00520 PRINT TAB(18); T(1); TAB(24); T(2); TAB(30); T(3)
00530 PRINT TAB(18); T(8); TAB(24);   0  ; TAB(30); T(4)
00540 PRINT TAB(18); T(7); TAB(24); T(6); TAB(30); T(5)
00550 RETURN
00560 :
00570 REM SUBROUTINE 2
00580 PRINT : INPUT "YOUR MOVE  "; P
00590 IF P < 1 OR P > 9 THEN 580
00600 LET T(P) = 11
00610 PRINT
00620 RETURN
```

26. FIFTEENS THE WINNER   On the table are playing cards with
the values of 1 - 9 (color and suit are irrelevant).   Each
player takes turns removing three cards.   The object of the
game is to keep the total value of one's own cards at exactly 15.   After picking up the three cards a player may exchange one for one of the three remaining on the table.   The
player to reach 15 first is the winner.   This is the way a
typical round might look:

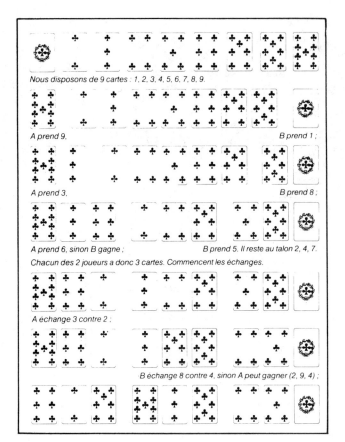

Nous disposons de 9 cartes : 1, 2, 3, 4, 5, 6, 7, 8, 9.

A prend 9.   B prend 1 ;

A prend 3.   B prend 8 ;

A prend 6, sinon B gagne ;   B prend 5. Il reste au talon 2, 4, 7.

Chacun des 2 joueurs a donc 3 cartes. Commencent les échanges.

A échange 3 contre 2 ;

B échange 8 contre 4, sinon A peut gagner (2, 9, 4) ;

2	9	4
7	5	3
6	1	8

Player A wins in this particular case.   Demonstrate that
this game is the equivalent of Tic-Tac-Toe.

(Note that the magic square [see above] is very important in this process. Now write the game program.

27. MOSER'S GAME You are presented with a list of words:

  HOT TANK TIED FORM HEAR BRIM WOES WASP SHIP
Two players take turns crossing out one word each (which
they mark with their names). The first one to cross out
three words that share exactly one letter in common is the
winner. Demonstrate that this is really a variation on
Tic-Tac-Toe. Write a program for the game.

206

28. JAM

Here is a diagram of a street map. A move in this game consists of blocking a street by filling in the whole street with your color. The first player to paint three streets that lead to the city with his own color is the winner. Believe it or not -- this is also a disguised version of Tic-Tac-Toe! Now write the program for it.

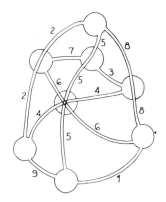

29. WILD TIC-TAC-TOE

This is played just like Tic-Tac-Toe -- with the variation that each player can play both O and X. Demonstrate that the player who goes first always wins.

In the reversed version of Wild Tic-Tac-Toe the player who completes a series of characters is the <u>loser</u>. Develop a game strategy and a program for this twist.

30. TRI-HEX

Instead of a 3x3 quadrant we have a playing field like the one pictured. The winner here is again the player who has placed three of his own marks in a row (on the same straight line). Demonstrate that the player who goes first can always win if he first puts his mark in the middle of one of the outside legs of the triangle. Use this to develop an optimal game program.

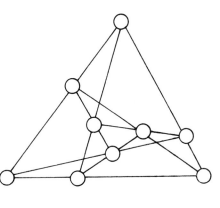

Example 6.5 BRIDGE BUILDER

```
       o   o   o   o   o
     x   x   x   x   x   x
       o   o   o   o   o   o
     x   x   x   x   x   x
       o   o   o   o   o   o
     x   x   x   x   x   x
       o   o   o   o   o   o
     x   x   x   x   x   x
       o   o   o   o   o   o
     x   x   x   x   x   x
       o   o   o   o   o
```

The field of play looks like the diagram pictured here (the number of lines or columns can be increased). Player A connects two adjacent circles either with a verticle or a horizontal line. The object is to make a complete line from the northern to the southern border of the field. Player B does the same thing with the x's. His goal is to make a connected line from east to west. Opponents' lines may not cross each other. The first to reach his goal is the winner.

The field of play looks like this after a typical round:

```
       o   o   o   o   o
     x——x   x   x   x   x
       o   o   o   o   o   o
     x   x   x——x——x——x
       o   o——o——o   o   o
     x   x——x——x   x   x
       o   o——o——o   o   o
     x   x——x——x   x   x
       o   o   o   o   o   o
     x   x   x   x——x   x
       o   o   o   o   o
```

Player B, moving from east to west, won this round. We need a program that will make this game possible on the screen and make sure that players move correctly. The program is not going to declare the winner. This is an interesting programming exercise that is left to the reader.

    A surprisingly simple game strategy was discovered not too long ago. It goes like this: "o opens with move shown below. Each time x crosses one of the dotted lines at one end, o crosses it at another.

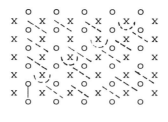

Try this strategy. You are going to win against every opponent but you're going to have to memorize carefully the position of the dotted lines beforehand. Here is the program:

```
00110 : PRINT CHR$(147)
00120 : PRINT "                    BRIDGE BUILDER
00130 : PRINT "                    --------------
00140 :
00150 REM TOPOLOGICAL GAME BY DAVID GALE
00160 :
00170 :
00180 REM *** MAIN PROGRAM ********************************
00185 :
00190 : GOSUB 320    : REM DIRECTIONS
00200 :
00210 : GOSUB 600    : REM INTIALIZATION
00220 :
00230 : GOSUB 780    : REM DISPLAY GAME AREA
00240 :
00250 : GOSUB 1110   : REM ORDER OF MOVES
00260 :
00270 : END
00280 :
00290 REM *** END OF MAIN PROGRAM *************************
00300 :
00310 :
00320 REM ### PROCEDURE 'DIRECTIONS' #######################
00330 :
00340 : PRINT
00350 : PRINT "DO YOU WANT TO SEE THE DIRECTIONS? (Y/N)
00360 :
00370 : GET A$ : IF A$ = "" THEN 370
00380 : IF A$ <> "Y" THEN RETURN
00390 :
00400 : PRINT CHR$(147)
00410 : PRINT "THIS IS A GAME FOR TWO PLAYERS. THE
00420 : PRINT "OBJECT IS TO CROSS A FIELD AND PREVENT THE
00425 : PRINT "OPPONENT FROM DOING THE SAME." : PRINT
00430 : PRINT "THE CHARACTERS USED ARE 'O' AND 'X'.
```

209

```
00440 : PRINT "'O' HAS TO TRY TO MOVE ACROSS THE FIELD FROM
00450 : PRINT "NORTH TO SOUTH. 'X' TRIES TO MOVE FROM WEST
00460 : PRINT "TO EAST. BOTH PLAYERS JUMP ONTO THE SQUARES
00465 : PRINT "MARKED FOR THEM." : PRINT : PRINT " WITH
00470 : PRINT "EACH JUMP A LINE IS DRAWN THROUGH THE SQUARE,
00480 : PRINT "MAKING THIS IMPOSSIBLE TO CROSS OVER AGAIN."
00490 : PRINT : PRINT "INPUT CONSISTS OF COORDINATES. THE
00500 : PRINT "DIRECTION IS DESIGNATED BY FIRST LETTERS OF
00505 : PRINT " THE COMPASS POINTS."
00510 :
00520 : PRINT : PRINT CHR$(18) "    *** HIT SPACE BAR. ***
00530 : GET A$ : IF A$ <> " " THEN 530
00540 :
00550 : RETURN
00560 :
00570 REM END OF 'DIRECTIONS' ############################
00580 :
00590 :
00600 REM ### PROCEDURE 'INITIALIZATION' #################
00610 :
00620 : LET H$(0) = "O"
00630 : LET H$(1) = "X"
00640 :
00650 : LET P(0) = 87 : REM CODE NUMBER FOR O
00660 : LET P(1) = 86 : REM CODE NUMBER FOR X
00670 :
00680 : LET B(0) = 64 : REM CODE FOR HORIZONTAL MOVE '-'
00690 : LET B(1) = 93 : REM CODE FOR PERPENDICULAR MOVE '|'
00700 :
00710 : LET SZ(0)= 40 : REM WINNING DIRECTION FOR O
00720 : LET SZ(1) = 1  : REM WINNING DIRECTION FOR X
00730 :
00740 : RETURN
00750 : REM END OF INITIALIZATION #########################
00760 :
00770 :
00780 : REM ### PROCEDURE 'DISPLAY PLAYING GRID' ##########
00790 :
00800 :      PRINT CHR$(147);
00810 :
00820 : REM --- O's
00830 :
00840 :   FOR I = 0 TO 5
00850 :     FOR J = 0 TO 4
00860 :       POKE 32812 + 160*I + 4*J, P(0)
00870 :     NEXT J
00880 :   NEXT I
00890 :
00900 : REM --- X's
00910 :
00920 :   FOR I = 0 TO 4
00930 :     FOR J = 0 TO 5
```

```
00940 :         POKE 32890 + 160*I + 4*J, P(1)
00950 :       NEXT J
00960 :    NEXT I
00970 :
00980 : REM --- LETTERING
00990 :
01000 :    FOR I = 75 TO 65 STEP -1
01010 :      PRINT CHR$(17) CHR$(18) CHR$(I)
01020 :    NEXT I
01030 :
01040 : PRINT CHR$(17) CHR$(18) "  A B C D E F G H I J K "
01050 :
01060 : RETURN
01070 :
01080 REM END OF PLAYING GRID DISPLAY ####################
01090 :
01100 :
01110 REM ### PROCEDURE 'ORDER OF MOVES' #################
01120 :
01130 : REM = SHORT PROCEDURE 'ORDER OF MOVES' =============
01140 :
01150 :    REM --- DETERMINING WHICH PLAYER MOVES
01160 :
01170 :       PRINT CHR$(19) TAB(24) CHR$(17) "WHO STARTS?
01180 :       PRINT TAB(24) CHR$(17) "O ... 0
01190 :       PRINT TAB(24) "X ... 1
01200 :
01210 :       GET A$ : IF A$ <> "0" AND A$ <> "1" THEN 1210
01220 :       LET W = VAL(A$)
01230 :
01240 :    REM --- ENTER SERIES OF MOVES
01250 :
01260 :       GOSUB 1500 : REM ERASE DISPLAY
01270 :       GOSUB 1610 : REM ENTER
01280 :
01290 :    REM --- INPUT CHECK
01300 :
01310 :       LET PK = 33610 + 2*X - 80*Y
01320 :
01330 :       IF PEEK(PK+4*V) <> P(W) AND V <> SZ(W) THEN 1240
01340 :       IF PEEK(PK+2*V) <> 32 THEN 1240
01350 :
01360 :    REM --- ENTERING INTO PLAYING GRID
01370 :
01380 :       FOR I = 1 TO 3
01390 :         POKE PK + V*I, B(C)
01400 :       NEXT I
01410 :
01420 :    REM --- NEXT PLAYER
01430 :
01440 :       LET W = 1 - W
01450 :
```

```
01460 :      GOTO 1240 : REM NEW MOVE
01470 :
01480 : REM END OF SHORT PROCEDURE 'ORDER OF MOVES' ========
01490 :
01500 : REM +++ SUBPROCEDURE 'ERASE' ++++++++++++++++++++++
01510 :
01520 :    PRINT CHR$(19)
01530 :    FOR I = 0 TO 13
01540 :      PRINT TAB(24) "                 "
01550 :    NEXT I
01560 :
01570 :    RETURN
01580 :
01590 : REM ++++ END OF ERASE ++++++++++++++++++++++++++++++
01600 :
01610 : REM +++ SUBPROCEDURE 'INPUT' ++++++++++++++++++++++
01620 :
01630 :    REM ANNOUNCE PLAYER NOW TAKING TURN
01640 :
01650 :      PRINT CHR$(19)
01660 :      PRINT TAB(24) "PLAYER NOW TAKING TURN : "
01670 :      PRINT TAB(24) CHR$(17) CHR$(18) " " H$(W) " "
01680 :
01690 :    REM ENTERING THE X-COORDINATES
01700 :
01710 :      PRINT TAB(24) CHR$(17) "X:= ";
01720 :      GET A$ : IF A$ = "" THEN 1720
01730 :      LET A = ASC(A$)
01740 :      IF A < 65 OR A > 75 THEN 1720
01745 :      LET T = 2 * INT(A/2) : LET T$ = "?" + CHR$(157)
01750 :      IF W = 0 AND A <> T THEN PRINT T$; : GOTO 1720
01760 :      IF W = 1 AND A =  T THEN PRINT T$; : GOTO 1720
01770 :      PRINT A$
01780 :      LET X = A - 65
01790 :
01800 :    REM ENTERING THE Y-COORDINATES
01810 :
01820 :      PRINT TAB(24) CHR$(17) "Y:= ";
01830 :      GET A$ : IF A$ = "" THEN 1830
01840 :      LET A = ASC(A$)
01850 :      IF A < 65 OR A > 75 THEN 1830
01855 :      LET T = 2 * INT(A/2)
01860 :      IF W = 0 AND A = T  THEN PRINT T$; : GOTO 1830
01870 :      IF W = 1 AND A <> T THEN PRINT T$; : GOTO 1830
01880 :      PRINT A$
01890 :      LET Y = A - 65
01900 :
01910 :    REM ENTERING THE DIRECTION
01920 :
01930 :      PRINT TAB(24) CHR$(17) CHR$(17) "DIRECTION:
01940 :      PRINT TAB(24) "(E,S,W,N)
01950 :      PRINT TAB(24) CHR$(17) "D:= ";
```

212

```
01960 :        GET D$ : PRINT D$
01970 :        IF D$ = "E" THEN V =    1 : C = 0 : GOTO 2030
01980 :        IF D$ = "S" THEN V =   40 : C = 1 : GOTO 2030
01990 :        IF D$ = "W" THEN V =   -1 : C = 0 : GOTO 2030
02000 :        IF D$ = "N" THEN V =  -40 : C = 1 : GOTO 2030
02010 :        GOTO 1960
02020 :
02030 :        RETURN
02035 :
02040 : REM END OF INPUT ++++++++++++++++++++++++++++++++++++++
02050 :
02060 REM END OF MOVE SERIES
02070 :
02080 :
```

ANOTHER BRAINTEASER:

31. The game of Hex was also ori-
ginated by Piet Hein. It's played
on a honeycomb -- hexagons joined
together. The number of hexagons
can vary (it is usually 11x11).
If you use 5x5 hexagons then the
game is called Mini-Hex.
The goal of the game is to make an

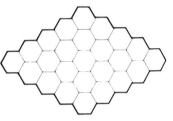

unbroken connection between two opposing sides. To do this
the players put down white or black playing pieces or they
mark the fields with their colors. Write a program for Hex
with the computer as the game master. No general game
strategy is known for this but it's easy to see that when
the field of play is small the player who opens always wins.
Make it possible for the computer to do this.

****

The pinnacle of all games of strategy is, of course,
chess. Because this game has traditionally been the favorite test of human intelligence, it has fascinated computer
experts more than any other game. Once it is possible to
teach a machine to play chess perfectly it will mean that
the word "intelligent" will finally apply to technology.

Artificial intelligence in the true sense of the word will then be a reality.

Long before the computer age people had been trying to invent chess-playing machines, or at least pretending they had been invented. There is an essay by Edgar Allen Poe called "Maelzel's Chess-Player" (1836) in which Poe describes a machine invented in 1769 by an Hungarian nobleman, the Baron von Kempelen. Maelzel toured the major cities of the world demonstrating the marvelous capabilities of von Kempelen's "automatic chess-player." Poe's attempt to compare the powers of the chess computer with the calculating machine made by the computer pioneer Babbage is interesting:

"Arithmetical or algebraical calculations are, from their very nature, fixed and determinate. Certain data being given, certain results necessarily and inevitably follow. These results have dependence upon nothing, and are influenced by nothing but the data originally given.....Upon starting it in accordance with the data of the question to be solved, it should continue its movements regularly, progressively, and undeviatingly towards the required solution....But the case is widely different with our Chess-Player. With him there is no determinate progression. No one move in chess necessarily follows upon any other....Now even granting...that the movements of the Automaton Chess Player were in themselves determinate, they would be necessarily interrupted and disarranged by the indeterminate will of this antagonist. There is then no analogy whatever between the operations of the Chess-Player and those of the calculating machine of Mr. Babbage, and if we choose to call the former a pure machine, we must be prepared to admit that it is, beyond all comparison, the most wonderful of the inventions of mankind."

*****

We need to develop a program to play chess. It isn't necessary that ours be so powerful that it can compete with the major league chess-playing programs. All we are looking for is something to demonstrate the basic ideas and methods behind such programs. It is up to you to dig deeper into the subject if you want.

214

Example 6.6 MINI-CHESS

You have a board that measures 6x6 squares. The object is to play a scaled-down game of chess on this board without using knights. The computer plays with the white pieces.

Let's represent the chess board as a two-dimensional table F(0..7, 0..7). The pieces are coded like this:

	white	black
pawn	1	-1
bishop	2	-2
rook	3	-3
queen	4	-4
king	5	-5

This is the way the starting position looks for the computer:

i \ j	0	1	2	3	4	5	6	7
0	10	10	10	10	10	10	10	10
1	10	3	2	4	5	2	3	10
2	10	1	1	1	1	1	1	10
3	10	0	0	0	0	0	0	10
4	10	0	0	0	0	0	0	10
5	10	-1	-1	-1	-1	-1	-1	10
6	10	-3	-2	-4	-5	-2	-3	10
7	10	10	10	10	10	10	10	10

The i coordinate indicates the row (horizontal line of squares) and the j indicates the column (vertical line). $F(1,4) = 5$ means that the white king is at the point of intersection of the first row and the fourth column. The 10's along the edge indicate the border of the field of play. This is how the board looks on the screen:

R	B	Q	K	B	R	1
P	P	P	P	P	P	2
						3
						4
p	p	p	p	p	p	5
r	b	q	k	b	r	6
F	E	D	C	B	A	

The capital letters represent the white pieces (the computer) and the small letters represent black (the player).

The entire program is organized into familiar sections: Directions, Initialization, Sequence of Moves, Exit or Repeat. The actual round of moves is divided into:

> Display board
> Computer move
> Display board
> Player move

This sequence repeats until the game is over. The process involved in the computer's move looks like this:

The computer examines every square (i,j) to look for its own (white) piece. The criterion is: F(i,j) > 0. If it finds this, then it next determines which piece is on the square (rook, bishop, etc.). Then it hypothetically tests all the possible moves this piece could make.

```
30   FOR  I = 1 TO 6
40     FOR  J = 1 TO 6
50       IF  F(I,J) < 0 THEN 70
60       ON  F(I,J)  GOSUB 100, 200, 300, 400, 500
70     NEXT J
80   NEXT I
```

The rest of the activity take place in the procedures for the various moves the pieces can make: pawns (100), bishops (200), rooks (300), queen (400), king (500). Before moving a piece, the computer has to determine whether the move is legal.

216

This involves making sure its own piece isn't already on that square or that the move won't endanger its own king. Moves like this and other illegal moves are interpreted as not permitted and will be skipped. Now, in order to select the best move permitted, the computer must analyze the playing position derived from the hypothetical test move. The computer takes the following points into account to evaluate its position:

(1) Is the piece threatened by an opponent's piece, and is it protected by one of its own pieces?

(2) Has the piece been moved to its best advantage so as to threaten as many squares as possible?

A further point is whether the piece can capture one of the opponent's pieces. Unfortunately, in a program of this limited scope we can't include this element of strategy as part of the move. To do so, one would have to examine other possible moves of the opponent. The evaluation function we're using at this point is "static."

The evaluation process first tests whether the planned move exposes the player's own king. If this proves to be the case, then the move is ignored and the computer skips to the next possibility which then tells the computer the "square value," or the effectiveness of the move. To do this the computer invokes the procedures "possible pawn moves," "possible rook moves," and "possible queen moves". These tell the computer how many free squares the piece can reach. The most difficult procedure to program is the one involving the material value of the move.

First, the value of the pieces protected by the piece we are preparing to move is entered into a table. Then, into another table go the values of the opponent's pieces that threaten our own. The value of the threatening pieces is subtracted from the value of the protected pieces (pawn: 10; bishop: 30; rook: 50; queen: 100) and the difference is added to the total value.

This process has to be repeated for each square (i,j). While doing so, the computer remembers each highest value and the move it went with. Then the computer executes the move having the highest total value. All these hypothetical test moves are displayed on the screen so the player can follow the process visually as the machine plans its game. (In the program, note that the use of LET is optional.)

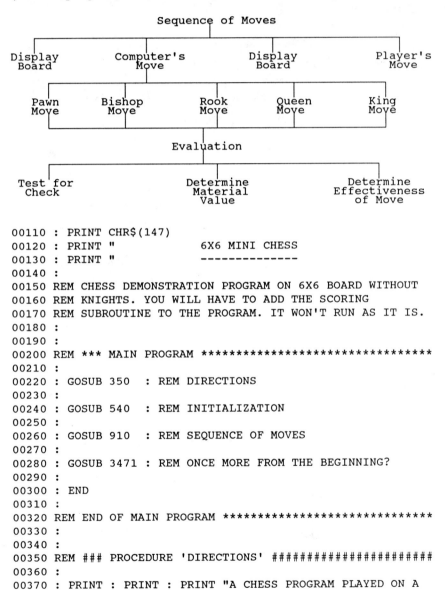

Sequence of Moves

Display Board   Computer's Move   Display Board   Player's Move

Pawn Move   Bishop Move   Rook Move   Queen Move   King Move

Evaluation

Test for Check   Determine Material Value   Determine Effectiveness of Move

```
00110 : PRINT CHR$(147)
00120 : PRINT "            6X6 MINI CHESS
00130 : PRINT "            --------------
00140 :
00150 REM CHESS DEMONSTRATION PROGRAM ON 6X6 BOARD WITHOUT
00160 REM KNIGHTS. YOU WILL HAVE TO ADD THE SCORING
00170 REM SUBROUTINE TO THE PROGRAM. IT WON'T RUN AS IT IS.
00180 :
00190 :
00200 REM *** MAIN PROGRAM ********************************
00210 :
00220 : GOSUB 350  : REM DIRECTIONS
00230 :
00240 : GOSUB 540  : REM INITIALIZATION
00250 :
00260 : GOSUB 910  : REM SEQUENCE OF MOVES
00270 :
00280 : GOSUB 3471 : REM ONCE MORE FROM THE BEGINNING?
00290 :
00300 : END
00310 :
00320 REM END OF MAIN PROGRAM ****************************
00330 :
00340 :
00350 REM ### PROCEDURE 'DIRECTIONS' #####################
00360 :
00370 : PRINT : PRINT : PRINT "A CHESS PROGRAM PLAYED ON A
```

```
00380 : PRINT "6X6 BOARD TO SPEED UP THE GAME. THE KNIGHTS
00390 : PRINT "HAVE BEEN REMOVED BUT OTHERWISE, NORMAL CHESS
00395 : PRINT "RULES APPLY. MOVES ARE ENTERED BY USING
00400 : PRINT "LANGER NOTATION, E.G, 'BA3-B2'. YOU PLAY THE
00400 : PRINT "BLACK CHESS PIECES (LOWER CASE LETTERS). THE
00410 : PRINT "COMPUTER PLAYS WHITE (UPPER CASE).
00420 :
00430 : PRINT : PRINT : PRINT CHR$(18) "   **HIT ANY KEY**
00440 :
00450 : GET T$ : IF T$ = "" THEN 450
00460 :
00470 : PRINT CHR$(147)
00480 :
00490 : RETURN
00500 :
00510 REM END OF DIRECTIONS ###############################
00520 :
00530 :
00540 REM ### PROCEDURE 'INITIALIZATION' ##################
00550 :
00560 : REM --- DATA
00570 :
00580 DATA 3,-3,2,-2,4,-4,5,-5,2,-2,3,-3 : REM FIGURE CODES
00590 DATA K,Q,R,B,P," ","l","L","l","-","'" : REM FIGURE
00595 :                                      REM CHARACTERS
00600 :
00610 : REM --- TABLES
00620 :
00630 :    DIM F(7,7) : REM FIELD OF PLAY
00640 :    DIM Z$(11) : REM FIGURE CHARACTERS
00650 :
00660 : REM --- SET UP THE PAWNS
00670 :
00680 :    FOR I = 1 TO 6 : F(2,I) = 1 : F(5,I) = -1 : NEXT I
00690 :
00700 : REM --- SET THE FIGURES
00710 :
00720 :    FOR I = 1 TO 6 : READ F(1,I) : READ F(6,I)
00725 :    NEXT I
00730 :
00740 : REM --- READ FIGURE CHARACTERS
00750 :
00760 :    FOR I = 1 TO 11 : READ Z$(I) : NEXT I
00770 :
00780 : REM --- SET LOWER CASE DISPLAY
00790 :
00800 :    POKE 59468,14
00810 :
00820 : REM --- CHARACTER SERIES TO MOVE CURSOR
00830 :
00835 :    LET O$ = CHR$(17)
00840 :    FOR I = 1 TO 4
```

```
00845 :      LET O$ = O$ + O$
00850 :    NEXT I
00855 :
00860 :    RETURN
00870 :
00880 :REM END OF INITIALIZATION #########################
00890 :
00900 :
00910 REM ### PROCEDURE 'MOVE SERIES' ####################
00920 :
00930 : REM = SHORT PROCEDURE 'MOVE SERIES' ================
00940 :
00950 :    GOSUB 1100 : REM DISPLAY BOARD
00960 :
00970 :    GOSUB 1270 : REM COMPUTER'S MOVE
00980 :
00990 :    GOSUB 1100 : REM DISPLAY BOARD
01000 :
01010 :    GOSUB 3080 : REM PLAYER'S MOVE
01020 :
01030 :    GOTO   950 : REM NEW MOVE
01040 :
01050 :    RETURN
01060 :
01070 : REM END OF SHORT PROCEDURE 'MOVE SERIES' ===========
01080 :
01090 :
01100 : REM +++ SUBPROCEDURE 'DISPLAY BOARD' +++++++++++++++
01110 :
01120 :    PRINT CHR$(19)
01130 :
01140 :    FOR I = 1 TO 6
01150 :      PRINT "┌─┐ ┌─┐ ┌─┐ ┌─┐ ┌─┐ ┌─┐"
01160 :      FOR J = 1 TO 6
01170 :        PRINT "│" Z$(F(I,J)+6) "│" ;
01180 :      NEXT J
01190 :      PRINT I
01200 :    NEXT I
01210 :    PRINT "└─┘ └─┘ └─┘ └─┘ └─┘ └─┘"
01220 :    PRINT " F   E   D   C   B   A  "
01230 :    PRINT : PRINT
01240 :
01250 :    RETURN
01260 :
01270 : REM +++ SUBPROCEDURE COMPUTER MOVE +++++++++++++++++
01280 :
01290 :    REM - SHORT PROCEDURE COMPUTER MOVE --------------
01300 :
01310 :      FOR I = 1 TO 6
01320 :        FOR J = 1 TO 6
01330 :          IF F(I,J) <= 0 THEN 1350
01340 :          ON F(I,J) GOSUB 1400, 1500, 1660, 1870, 1960
```

```
01350 :       NEXT J
01360 :       NEXT I
01370 :       RETURN
01375 :
01380 :   REM END OF SHORT PROCEDURE 'COMPUTER MOVE' -------
01390 :
01400 :   REM +++ SUBPROCEDURE 'PAWNS' MOVES' ++++++++++++++
01410 :
01420 :       IF F(I+1,J) = 0 THEN X=I+1 : Y=J    : GOSUB 2080
01430 :       IF F(I+1,J+1)<0 THEN X=I+1 : Y=J+1 : GOSUB 2080
01440 :       IF F(I+1,J-1)<0 THEN X=I+1 : Y=J-1 : GOSUB 2080
01450 :
01460 :       RETURN
01470 :
01480 :   REM END OF PAWNS' MOVES ++++++++++++++++++++++++++
01490 :
01500 :   REM +++ SUBPROCEDURE 'BISHOPS' MOVES' +++++++++++++
01510 :
01520 :       FOR K1 = -1 TO 1 STEP 2
01530 :         FOR K2 = -1 TO 1 STEP 2
01540 :           LET X = I : LET Y = J
01550 :           LET X = X + K1 : LET Y = Y + K2
01560 :           IF F(X,Y) < 0 THEN GOSUB 2080 : GOTO 1590
01570 :           IF F(X,Y) > 0 THEN 1590 : REM OWN MAN
01580 :           GOSUB 2080 : GOTO 1550 : REM SCORE AND
01585 :                                    REM CONTINUE
01590 :         NEXT K2
01600 :       NEXT K1
01610 :
01620 :       RETURN
01630 :
01640 :   REM END OF BISHOPS' MOVES +++++++++++++++++++++++++
01650 :
01660 :   REM +++ SUBPROCEDURE ' ROOKS' MOVES ' +++++++++++++
01670 :
01680 :       LET Y = J
01690 :       FOR K = -1 TO 1 STEP 2
01700 :         LET X = I
01710 :         LET X = X + K
01720 :         IF F(X,Y) > 0 THEN 1750
01730 :         IF F(X,Y) < 0 THEN GOSUB 2080 : GOTO 1750
01740 :         GOSUB 2080 : GOTO 1710
01750 :       NEXT K
01760 :
01770 :       LET X = I
01780 :       FOR K = -1 TO 1 STEP 2
01790 :         LET Y = Y + K
01800 :         IF F(X,Y) > 0 THEN 1830
01810 :         IF F(X,Y) < 0 THEN GOSUB 2080 : GOTO 1830
01820 :         GOSUB 2080 : GOTO 1790
01830 :       NEXT K
01840 :
```

221

```
01850 :      RETURN
01860 :
01870 :    REM +++ SUBPROCEDURE ' QUEENS' MOVES ' ++++++++++
01880 :
01890 :      GOSUB 1500 : REM BISHOPS' MOVES
01900 :      GOSUB 1660 : REM ROOKS' MOVES
01910 :
01920 :      RETURN
01930 :
01940 :    REM END OF QUEENS' MOVES ++++++++++++++++++++++++
01950 :
01960 :    REM +++ KING'S MOVES ++++++++++++++++++++++++++++
01970 :
01980 :      FOR K1 = -1 TO 1
01990 :        FOR K2 = -1 TO 1
02000 :          IF F(I+K1,J+K2) >= 1 THEN 2010
02005 :          X = I + K1 : Y = J + K2 : GOSUB 2080
02010 :        NEXT K2
02020 :      NEXT K1
02030 :
02040 :      RETURN
02050 :
02060 :    REM END OF KINGS' MOVES ++++++++++++++++++++++++++
02070 :
02080 :    REM +++ SUBPROCEDURE 'EVALUATION' +++++++++++++++
02090 :
02100 :      REM - SHORT PROCEDURE 'SCORE' ------------------
02110 :
02120 :        LET H1 = 32849 + (X-1)*80 + (Y-1)*3
02125 :          REM POSITION OF MOVE
02130 :        LET H2 = 32849 + (I-1)*80 + (J-1)*3
02135 :          REM POSITION OF PIECE
02140 :        H3 = PEEK(H1) : POKE H1,PEEK(H2) : POKE H2,32
02150 :
02160 :        REM DETERMINING MATERIAL VALUE AND EFFECT OF
02170 :        REM MOVE MUST BE ENTERED HERE.
02180 :
02190 :        RETURN
02200 :
02210 :      REM END OF SHORT PROCEDURE 'SCORE' -------------
02220 :
02230 :      REM +++SUBPROCEDURE 'POSSIBLE BISHOPS' MOVES'+++
02240 :
02250 :        REM MUST BE ENTERED HERE
02260 :
02270 :        RETURN
02280 :
02290 :      REM END OF POSSIBLE BISHOPS' MOVES ++++++++++++
02300 :
02310 :      REM +++ SUBPROCEDURE 'POSSIBLE ROOKS' MOVES'++++
02320 :
02330 :        REM MUST BE ENTERED
```

222

```
02340 :
02350 :          RETURN
02360 :
02370 :    REM END OF POSSIBLE ROOKS' MOVES ++++++++++++++
02380 :
02390 :    REM SUBPROCEDURE 'POSSIBLE QUEENS' MOVES' ++++++
02400 :
02410 :       GOSUB 2230        : REM POSSIBLE BISHOPS' MOVES
02420 :       GOSUB 2310        : REM POSSIBLE ROOKS' MOVES
02430 :
02440 :       RETURN
02450 :
02460 :    REM END OF POSSIBLE QUEENS' MOVES ++++++++++++++
02470 :
02480 :    REM +++ SUBPROCEDURE 'TESTING FOR CHECK' +++++++
02490 :
02500 :       FOR II = 1 TO 6
02510 :         FOR JJ = 1 TO 6
02520 :           IF F(II,JJ) = 5 THEN KX = II : KY = JJ
02530 :         NEXT JJ
02540 :       NEXT II
02550 :
02560 :       LET H$ = "CHECK"
02570 :
02580 :       IF F(KX+1,KY+1) = -1 THEN RETURN
02590 :       IF F(KX-1,KY+1) = -1 THEN RETURN
02600 :
02610 :       LET U = KX  : LET V = KY
02620 :       LET U = U+1 : LET V = V+1
02625 :       IF F(U,V) = 0 THEN 2620
02630 :       IF F(U,V) = -2 THEN RETURN : REM BISHOP
02635 :                                   REM THREATENS CHECK
02640 :       IF F(U,V) = -4 THEN RETURN: REM QUEEN
02645 :                                   REM THREATENS CHECK
02650 :
02660 :       LET U = KX  : LET V = KY
02670 :       LET U = U+1 : LET V = V-1
02675 :       IF F(U,V) = 0 THEN 2670
02680 :       IF F(U,V) = -2 OR F(U,V) = -4 THEN RETURN
02690 :
02700 :
02710 :       LET U = KX  : LET V = KY
02720 :       LET U = U-1 : LET V = V+1
02725 :       IF F(U,V) = 0 THEN 2720
02730 :       IF F(U,V) = -2 OR F(U,V) = -4 THEN RETURN
02740 :
02750 :       LET U = KX  : LET V = KY
02760 :       LET U = U-1 : LET V = V-1
02765 :       IF F(U,V) = 0 THEN 2760
02770 :       IF F(U,V) = -2 OR F(U,V) = -4 THEN RETURN
02780 :
02790 :       LET U = KX  : LET V = KY
```

223

```
02800 :        LET U = U+1 : IF F(U,V) = 0 THEN 2800
02810 :        IF F(U,V) = -3 THEN RETURN : REM ROOK
02815 :                                    REM THREATENS CHECK
02820 :        IF F(U,V) = -4 THEN RETURN : REM QUEEN
02825 :                                    REM THREATENS CHECK
02830 :
02840 :        LET U = KX
02850 :        LET U = U-1 : IF F(U,V) = 0 THEN 2850
02860 :        IF F(U,V) = -3 OR F(U,V) = -4 THEN RETURN
02870 :
02880 :        LET V = KY
02890 :        LET V = V+1 : IF F(U,V) = 0 THEN 2890
02900 :        IF F(U,V) = -3 OR F(U,V) = -4 THEN RETURN
02910 :
02920 :        LET V = KY
02930 :        LET V = V-1 : IF F(U,V) = 0 THEN 2930
02940 :        IF F(U,V) = -3 OR F(U,V) = -4 THEN RETURN
02950 :
02960 :        FOR M = -1 TO 1
02970 :          FOR N = -1 TO 1
02980 :            IF F(KX+M,KY+N) = -5 THEN RETURN
02990 :          NEXT N
03000 :        NEXT M
03010 :
03020 :        LET H$ = "NOT CHECK"
03030 :
03040 :        RETURN
03050 :
03060 : REM END OF COMPUTER'S MOVE ++++++++++++++++++++++++++
03070 :
03080 : REM SUBPROCEDURE 'PLAYER'S MOVE' ++++++++++++++++++++
03090 :
03100 : REM --- INPUT
03110 :
03120 :    PRINT CHR$(19);O$;"
03130 :    PRINT "
03140 :    PRINT CHR$(145) CHR$(145);
03145 :    INPUT "YOUR MOVE: "; Z$
03150 :
03160 :    IF LEN(Z$) <> 6 THEN 3340 : REM CATCH ERROR
03170 :    IF MID$(Z$,4,1) <> "-" THEN 3340 : REM CATCH ERROR
03180 :
03190 : REM --- ENTER ON BOARD
03200 :
03210 :    LET VX = 7 - (ASC(MID$(Z$,2,1)) - 64)
03220 :    LET VY = VAL(MID$(Z$,3,1))
03230 :
03240 :    LET NX = 7 - (ASC(MID$(Z$,5,1)) - 64)
03250 :    LET NY = VAL(MID$(Z$,6,1))
03260 :
03270 :    IF Z$(F(VY,VX)+6) <> LEFT$(Z$,1) THEN 3340
03275 :    IF F(NY,NX) < 0 THEN 3340
```

224

```
03280 :
03290 :    LET F(NY,NX) = F(VY,VX)
03300 :    LET F(VY,VX) = 0
03310 :
03320 :    RETURN : REM BACK TO SHORT PROCEDURE 'MOVE SERIES'
03330 :
03340 : REM --- CATCH ERROR
03350 :
03360 :      PRINT CHR$(145)
03370 :      PRINT CHR$(145) CHR$(18) "         ERROR!
03380 :
03390 :      FOR I = 1 TO 1000 : NEXT I
03400 :
03410 :      GOTO 3120
03420 :
03430 : REM END OF PLAYER'S MOVE +++++++++++++++++++++++++++++
03440 :
03450 REM END OF MOVE SERIES ##############################
03460 :
03470 :
```

32. Add a section to the Mini Chess game that will score a player's moves.

33. Design a chess program for a board measuring 5x5. Use the following pieces: bishop, knight, rook, queen, king and pawns).

34. Expand the 6x6 chess program to the size of 8x8. Include two knights.

35. BLOCK TURNING   This game is played on a 3x3 board using markers having an X on one side an O on the other.   The starting position looks like this:

              X O X
              O X O
              X O X

A move consists of turning over all the stones of a quadrant ("block").   The quadrants have these dimensions: 1x1, 1x2, 1x3, 2x2, 2x3, 3x3.   A block can only be turned over when the upper left corner shows a stone with an X.   (Both players are seated on the same side of the board so there is no confusion about the location of this important corner.) Whoever is successful in turning all the stones to the O sides is the winner.   Players alternate taking turns.   A stone must be turned over every move.

     Develop a winning strategy by working recursively from the final position.

36. EUCLID'S GAME   We have a rectangle with sides measured in whole numbers.   Two opponents play against each other this way: a move consists of cutting a square of any dimension off of the rectangle.   The catch is that the shape that remains must still be a rectangle.   The loser is the player who can't move any more because there is no more rectangle

to cut up. (The winner is the player who is able to divide the rectangle completely into squares.) Develop a strategy and program it yourself.

37. Here is a diagram for a game of Cat and Mouse. At the starting position the cat is on the spot marked 2 and the mouse is on 15. The cat goes first and tries to catch the mouse, that is, to occupy the same spot as the opponent. Only moves to adjacent spots via the curved lines are allowed. Now try to develop a game strategy.

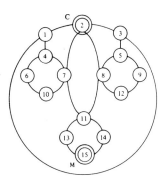

38. Two knights are on a chess board. One is trying to capture the other. The object is to threaten the square the opponent is about to jump to. In the beginning the knights are placed at opposite corners of the boards. Find a winning strategy and a program.

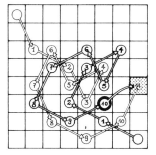

227

Here is a neat little game to close this chapter:

Counterexample 6.7   FOUR'S THE WINNER

```
10 DIMRR(11)
21 DIMRE(10,10),A3(2,100),12(2,100)
30 PRINTCHR$(147)
35 FORI=1TO10:PRINTCHR$(17);:NEXTI
40 PRINTCHR$(18)"                                    "
42 PRINTCHR$(18)"        FOUR IS THE WINNER          "
44 PRINTCHR$(18)"                                    "
50 FORZ=1TO1000:NEXTZ
55 PRINT:PRINT:PRINT:PRINT:PRINT
60 PRINT"DO YOU KNOW THE RULES OF THE GAME?"
70 GETZ$:IFZ$<>"Y"AND Z$<>"N"THEN70
80 IFZ$="N"THENGOSUB50000
100 PRINTCHR$(147)
110 FORZ=1TO400:NEXTZ
120 PRINT"  |   |   |   |   |   |   |   |   |   |   "
140 PRINT"  |---|---|---|---|---|---|---|---|---|   "
150 PRINT"  |   |   |   |   |   |   |   |   |   |   "
170 PRINT"  |---|---|---|---|---|---|---|---|---|   "
180 PRINT"  |   |   |   |   |   |   |   |   |   |   "
200 PRINT"  |---|---|---|---|---|---|---|---|---|   "
210 PRINT"  |   |   |   |   |   |   |   |   |   |   "
230 PRINT"  |---|---|---|---|---|---|---|---|---|   "
240 PRINT"  |   |   |   |   |   |   |   |   |   |   "
260 PRINT"  |---|---|---|---|---|---|---|---|---|   "
270 PRINT"  |   |   |   |   |   |   |   |   |   |   "
290 PRINT"  |---|---|---|---|---|---|---|---|---|   "
300 PRINT"  |   |   |   |   |   |   |   |   |   |   "
320 PRINT"  |---|---|---|---|---|---|---|---|---|   "
340 PRINT"  |   |   |   |   |   |   |   |   |   |   "
360 PRINT"  |---|---|---|---|---|---|---|---|---|   "
380 PRINT"  |   |   |   |   |   |   |   |   |   |   "
400 PRINT"  |---|---|---|---|---|---|---|---|---|   "
420 PRINT"  |   |   |   |   |   |   |   |   |   |   "
450 PRINT"  |___|___|___|___|___|___|___|___|___|   "
470 PRINT"   1   2   3   4   5   6   7   8   9"
475 PRINT:PRINT" READY TO BEGIN ?"
480 GETC$:IFC$<>"N"ANDC$<>"Y"THEN480
490 PRINTCHR$(145)"                            "CHR$(145)
495 IFC$="N"THENPRINT:GOTO 800
500 PRINT"    PLEASE BEGIN !"
600 GETA$
610 IFA$<>"1"ANDA$<>"2"ANDA$<>"3"ANDA$<>"4"ANDA$<>"5"THEN630
620 GOTO700
630 IFA$<>"6"ANDA$<>"7"ANDA$<>"8"ANDA$<>"9"THEN600
700 A=VAL(A$)
```

```
710 IFRR(A)>0THEN IFLOG(RR(A))/LOG(10)>9THEN600
720 SP=1
750 GOSUB5400:GOSUB5000
780 PRINTCHR$(145)"                              "
790 GOSUB12000:GOSUB20000 : IFSG$="WIN"THEN900
800 GOSUB2000
810 IFSG$="LOSE"THEN850
820 PRINTCHR$(145)"  YOUR MOVE   !       "
830 POKE158,0:GOTO600
850 PRINT" I'M AFRAID YOU LOSE.":GOTO910
900 PRINT" YOU WIN !"
910 FORY=1TO200:GETA$:NEXTY
920 GETA$:IFA$=""THEN920
999 :  RUN
2000 REM --- COMPUTER MOVES ---
2080 A=0
2090 SP=2:IFM(SP)=0THEN3050
2100 FORR=1TOM(SP)
2120 B=INT(A3(SP,R)):C=(A3(SP,R)-B)*10
2125 GOSUB30000
2130 IFC=0THEN3000
2140 IFRE(B,C-1)<>0THEN 3000
2200 NEXTR :GOTO3050
3000 A=B:IFA>0THEN3500
3050 SP=1:IFM(SP)=0THEN3150
3060 FORR=1TOM(SP)
3070 B=INT(A3(SP,R)):C=(A3(SP,R)-B)*10
3075 GOSUB30000
3080 IFC=0ANDB<>0THEN3900
3085 IFC=0THEN3100
3090 IFRE(B,C-1)<>0THEN3900
3100 NEXTR
3150 SP=2:IFN(SP)=0THENA=0:GOTO3285
3160 FORR=1TON(SP)
3170 B=INT(A2(SP,R)):C=(A2(SP,R)-B)*10
3175 GOSUB30000
3180 IFC=0THEN3250
3190 IFRE(B,C-1)<>0THEN3250
3200 GOTO3280
3250 GOSUB40000
3265 IFDF$="NO"THEN3280
3270 IFB<>0THEN3900
3280 NEXTR
3285 SP=1:IFN(1)<2THEN3305
3287 FORR=N(SP)-1TO1STEP-1
3289 B=INT(A2(SP,R)):C=(A2(SP,R)-B)*10
3291 GOSUB30000
3293 IFC=0THEN3297
3294 IFRE(B,C-1)<>0THEN3297
3296 GOTO3301
3297 GOSUB40000
3299 IFDF$="NO"THEN3301
```

229

```
3300 IFB<>0GOTO3900
3301 NEXTR
3305 IFA<>0THENB=A
3310 IFA=0THENB=INT(RND(1)*9+1)
3313 IFRR(B)=0THENC=0:GOTO3320
3315 C=INT(LOG(RR(B))/LOG(10))+2:IFC>9THENA=0:GOTO3310
3320 GOSUB40000
3330 IFDF$="NO"THENA=0:GOTO 3310
3400 GOTO3900
3500 SP=2:GOSUB5400:SG$="LOSE":GOTO4000
3900 A=B:SP=2:GOSUB5400:GOSUB12000
4000 GOSUB20000:GOSUB5000: RETURN
5000 REM --- MOVING ---
5010 ZE=86
5020 IFSP=2THENZE=ZE-5
5050 LE=32768
5060 FORU=0TOA*4-2
5070 POKELE+U,ZE
5080 FORYY=1TO20:NEXTYY : REM --HESITATION--
5090 POKELE+U,32
5100 NEXTU
5200 U=U+40
5210 ZV=PEEK(LE+U)
5220 IFPEEK(LE+U+40)=81ORPEEK(LE+U+40)=86THEN5350
5230 POKELE+U,ZE
5240 FORYY=1TO20:NEXTYY : REM--HESITATION--
5250 IFU>760THEN5380
5260 POKELE+U,ZV
5280 GOTO5200
5350 POKELE+U-40,ZE
5380 RETURN
5400 IFRR(A)=0THENRR(A)=SP:GOTO5450
5410 RR(A)=VAL(STR$(RR(A))+STR$(SP))
5450 LO=LEN(STR$(RR(A)))-2
5460 RE(A,LO)=SP
5500 RETURN
12000 IFSP=2THENZW=-2
12003 ZW=-ZW
12005 D=0:I=1 :FORQ=0TO3
12010 B=A:C=LO+1-Q
12013 IFB<1ORC<0ORB>9ORC>9THEN12190
12015 Y=RE(B,C)
12020 IFQ=0THENAX(I)=B+C/10:I=I+1
12030 IFY=SPTHEND=D+1
12040 IFY<>SPANDY<>0THEN12100
12050 GOTO12190
12100 IFD<>3THEN12120
12110 FORT=1TOI-1:M(SP)=M(SP)+1:A3(SP,M(SP))=AX(T)
12115 GOSUB15000:NEXTT
12120 IFD<>ZWTHEN12140
12130 FORT=1TOI-1:N(SP)=N(SP)+1:A2(SP,N(SP))=AX(T)
12135 GOSUB16000:NEXTT
```

230

```
12140 IFD=4THENSG$="WIN":GOTO13000
12180 IFQ=4THEN12200
12190 NEXTQ
12195 GOTO12100
12200 GOTO12400
12202 FORW=0TO3
12205 D=0:I=1 :FORQ=0TO3
12210 B=A-3+W+Q:C=LO
12213 IFB<1ORC<0ORB>9ORC>9THEN12380
12215 Y=RE(B,C)
12220 IFY=0THENAX(I)=B+C/10:I=I+1
12230 IFY=SPTHEND=D+1
12240 IFY<>SPANDY<>0THEN12300
12250 GOTO12380
12300 IFD<>3THEN12320
12310 FORT=1TOI-1:M(SP)=M(SP)+1:A3(SP,M(SP))=AX(T)
12315 GOSUB15000:NEXTT
12320 IFD<>2THEN12340
12330 FORT=1TOI-1:N(SP)=N(SP)+1:A2(SP,N(SP))=AX(T)
12335 GOSUB16000:NEXTT
12340 IFD=4THENSG$="WIN":GOTO13000
12350 D=0:I=1
12360 IFQ=4THEN12390
12380 NEXTQ
12385 GOTO12300
12390 NEXTW
12399 GOTO13000
12400 FORW=0TO3
12405 D=0:I=1 :FORQ=0TO3
12410 B=A-3+W+Q:C=LO+3-Q-W
12411 GOSUB30000
12413 IFB<1ORC<0ORB>9ORC>9THEN12580
12415 Y=RE(B,C)
12420 IFY=0THENAX(I)=B+C/10:I=I+1
12430 IFY=SPTHEND=D+1
12440 IFY<>SPANDY<>0THEN12500
12450 GOTO12580
12500 IFD<>3THEN12520
12510 FORT=1TOI-1:M(SP)=M(SP)+1:A3(SP,M(SP))=AX(T)
12515 GOSUB15000:NEXTT
12520 IFD<>ZWTHEN12540
12530 FORT=1TOI-1:N(SP)=N(SP)+1:A2(SP,N(SP))=AX(T)
12535 GOSUB16000:NEXTT
12540 IFD=4THENSG$="WIN":GOTO13000
12550 D=0:I=1
12560 IFQ=4THEN12590
12580 NEXTQ
12585 GOTO12500
12590 NEXTW
12600 FORW=0TO3
12605 D=0:I=1 :FORQ=0TO3
12610 B=A-3+W+Q:C=LO-3+Q+W
```

231

```
12611 GOSUB30000
12613 IFB<1ORC<0ORB>9ORC>0THEN12780
12615 Y=RE(B,C)
12620 IFY=0THENAX(I)=B+C/10:I=I+1
12630 IFY=SPTHEND=D+1
12640 IFY<>SPANDY<>0THEN12700
12650 GOTO12780
12700 IFD<>3THEN12720
12710 FORT=1TOI-1:M(SP)=M(SP)=1:A3(SP,M(SP))=AX(T)
12715 GOSUB15000:NEXTT
12720 IFD<>ZWTHEN12740
12730 FORT=1TOI-1:N(SP)=N(SP)+1:A2(SP,N(SP))=AX(T)
12735 GOSUB16000:NEXTT
12740 IFD=4THENSG$="WIN":GOTO13000
12750 D=0:I=1
12760 IFQ=4THEN12790
12780 NEXTQ
12785 GOTO12700
12790 NEXTW
12800 GOTO12202
13000 RETURN
15000 :
15003 IFI=1THENM(SP)=M(SP)-1:GOTO15500
15008 IFM(SP)=1THEN15070
15009 FORR=1TOM(SP)-1
15010 IFA3(SP,R)=A3(SP,M(SP))THEN15050
15020 NEXTR
15030 N=INT(A3(SP,M(SP))):M=INT((A3(SP,M(SP))-N)*10+.2)
15040 IFRE(N,M)<>0THENM(SP)=M(SP)-1:GOTO15500
15045 GOTO15070
15050 M(SP)=M(SP)-1:A3(SP,M(SP)+1)=0:GOTO15500
15070 FORR=1TON(SP)
15080 IFA2(SP,R)=A3(SP,M(SP))THEN15100
15090 NEXTR:GOTO15500
15100 N(SP)=N(SP)-1:FORE=RTON(SP)
15110 A2(SP,E)=A2(SP,E+1)
15120 NEXTE:A2(SP,N(SP)+1)=0
15500 RETURN
16000 IFI=1THENN(SP)=N(SP)-1:GOTO16500
16002 IFN(SP)=1THEN16200
16004 N=INT(A2(SP,N(SP))):M=INT(A2(SP,N(SP))-N+.2)
16006 IFRE(N,M)<>0THEN16100
16010 FORR=1TON(SP)-1
16020 IFA2(SP,N(SP))=A2(SP,R)THEN16100
16040 NEXTR
16060 GOTO16200
16100 N(SP)=N(SP)-1:A2(SP,N(SP)+1)=0
16150 GOTO16500
16200 FORR=1TOM(SP)
16220 IFA2(SP,N(SP))=A3(SP,R)THEN16300
16240 NEXTR
16260 GOTO16500
```

```
16300 N(SP)=N(SP)-1:A2(SP,N(SP)+1)=0
16500 RETURN
20000 ZX=A+LO/10
20020 FORR=1TOM(1)
20040 IFA3(1,R)=ZXTHEN20080
20060 NEXTR:GOTO20200
20080 FORE=RTOM(1)-1
20100 A3(1,E)=A3(1,E+1)
20120 NEXTE:A3(1,M(1))=0: M(1)=M(1)-1
20200 FORR=1TON(1)
20240 IFA2(1,R)=ZXTHEN20280
20260 NEXTR:GOTO20400
20280 FORE=RTON(1)-1
20300 A2(1,E)=A2(1,E+1)
20320 NEXTE:A2(1,N(1))=0: N(1)=N(1)-1
20400 FORR=1TON(2)
20440 IFA2(2,R)=ZXTHEN20480
20460 NEXTR:GOTO20600
20480 FORE=RTON(2)-1
20500 A2(2,E)=A2(2,E+1)
20520 NEXTE:A2(2,N(2))=0: N(2)=N(2)-1
20600 FORR=1TOM(2)-1
20640 IFA3(2,R)=ZXTHEN20680
20660 NEXTR:GOTO21000
20680 FORE=RTOM(2)-1
20700 A3(2,E)=A3(2,E+1)
20720 NEXTE:A3(2,M(2))=0: M(2)=M(2)-1
21000 RETURN
30000 C=INT(C+.2):RETURN
40000 REM ---NO WIN FOR OPPONENT---
40050 DF$=""
40070 IFRR(B)=0THENC=1:GOTO40200
40100 C=INT(LOG(RR(B))/LOG(10))+2
40200 FORF=1TOM(1)
40300 IFA3(1,F)=B+C/10THENDF$="NO"
40400 NEXTF
40500 IFC>9THENDF$="NO"
41000 RETURN
50000 PRINTCHR$(147):PRINT:PRINT
50005 PRINTCHR$(18)"                                  "
50010 PRINTCHR$(18)"   FOUR'S THE WINNER  -   RULES   "
50015 PRINTCHR$(18)"                                  "
50030 PRINT:PRINT:PRINT:PRINT
50050 PRINT"THE BOARD IS DISPLAYED WHICH CONSISTS OF 9
50060 PRINT"NUMBERED COLUMNS. YOU AND THE COMPUTER TAKE
50070 PRINT"TURNS DROPPING 'STONES' ONTO THESE. YOUR OBJECT
50080 PRINT"IS TO TRY TO BE THE FIRST TO GET 4 STONES INTO
50090 PRINT"ONE ROW."
50150 GETZ$:IFZ$=""THEN50150
50200 PRINTCHR$(147):PRINT:PRINT
50220 PRINT"THIS ROW CAN BE DIAGONAL, HORIZONTAL OR
50250 PRINT"VERTICAL":PRINT:PRINT
```

```
50300 PRINT"YOUR PLAYING STONE:    'X'"
50310 PRINT
50320 PRINT"THE COMPUTER'S STONE: 'O'"
50330 PRINT:PRINT"YOUR STONE FALLS INTO A COLUMN AS SOON"
50340 PRINT"AS YOU PRESS ITS CORRESPONDING NUMBER."
50350 FORZZ=1TO5:PRINT:NEXTZZ
50500 PRINT"   GOOD    LUCK !!"
50600 GETZ$:IFZ$=""THEN50600
52000 RETURN
```

Enter the program and play around with it. You will find it
both fun and suspenseful. But please don't expect to understand the program right away. It is included as a "counterexample" to demonstrate the _wrong_ way to do things and to
emphasize that the programs you learn from this book are
supposed to be

UNDERSTANDABLE!

To program clearly and understandably means to program
in a structured fashion -- like the models you have encoun-
tered so far in this book. When you think you have finished
a nice program, then take _twice_ the amount of time again to
rewrite it.

Say to yourself, "Now I'm going to rewrite this so that
I can understand it again in a month and so anybody else can
understand it immediately." Sounds like hard work? It is.
But without it your nice program is worthless.

CHAPTER VII

VARIED STRATEGIES

In this chapter we are going to get to know games that belong to the group using so-called varied strategies. These are different from games of pure luck (chapter 5) but they aren't true games of strategy either (as in chapter 6). They represent an in-between type. Not only chance but also players' decisions influence the results. And yet the role of chance here is different. Whereas it affected those games of pure luck as a sort of "finger of fate" that tickled the unsuspecting player, here chance is constructively included in the strategy.

Example 7.1 CHOOSE UP FOR PENNIES

Player A (Alex) and Player B (Beth) are playing by the following rules: they simultaneously "throw" one or two fingers. Alex gives Beth as many pennies as fingers are thrown when this number is even. But when the number of fingers showing is odd, then Beth pays Alex that number of pennies.

The computer plays Alex. Our object is to program him to play as intelligently as possible.

The following four situations are possible:

Both throw one finger: Alex pays Beth 2 pennies.
Alex throws one, Beth two fingers: Beth pays Alex 3 pennies.
Alex throws two fingers, Beth one: Beth pays Alex 3 pennies.
Both throw 2 fingers: Alex pays Beth 4 pennies.

We can express these possibilities with a simple table:

		Beth	
F		1	2
A l e x	1	-2	3
	2	3	-4

The game looks fair because the total of payments Alex makes to Beth is equal to those Beth makes to Alex. But how should one play the game in order to gain an advantage? Alex will soon figure out that he can't keep showing only one finger, because as soon as Beth sees this, that's all she will throw. This would mean a loss of two pennies with each throw. But for that matter, Beth won't keep throwing the same finger because Alex will soon develop a strategy to counter it.

Alex is going to have to alternate between showing one finger and showing two. Like this: 1212121212..., or like this: 112112112112, or...? But Beth is eventually going to figure this out too. This means that there must not be an obvious pattern to Alex's finger throwing. It has to be random. This is best achieved by letting the number of "1's" and "2's" be determined by chance. A roulette wheel could guide him if he spun it before each move. Beth is going to proceed similarly. The question is: how do you divide up the segments of the circle that make up this wheel? That is, what is the mathematical relationship between the number of 1's and 2's? We assume there are a few more 1's than 2's because that keeps the losses smaller. Here is a game simulation to figure this out.

Simulation of Choose Up for Pennies

```
How many moves? 100
Frequency with which Player A shows one finger: 0.75
Frequency with which Player B shows one finger: 1
Average profit for player A per round: -0.6

How many moves? 100
Frequency with which Player A shows one finger: 0.5
Frequency with which Player B shows one finger: 0
Average profit for player A per round: -0.64

How many moves? 500
Frequency with which Player A shows one finger: 0.6
Frequency with which Player B shows one finger: 0.5
Average profit for player A per round: 0.11
```

And here is the program:

236

```
00110 : PRINT CHR$(147)
00120 : PRINT "  SIMULATION OF CHOOSE UP FOR PENNIES
00130 : PRINT "  -----------------------------------
00140 :
00150 REM THE COMPUTER PLAYS N ROUNDS OF PENNY TOSS TO
00160 REM DETERMINE THE OPTIMAL COMBINATION OF 1 AND 2.
00170 :
00180 REM === INPUT AND BEGINNING VALUES ===================
00190 :
00200 : PRINT
00210 : INPUT "HOW MANY ROUNDS "; N
00220 :
00230 : PRINT "FREQUENCY PLAYER A SHOWS 1 FINGER "
00235 : INPUT "(NUMBER BETWEEN 0 AND 1!)"; PA
00240 : IF PA < 0 OR PA > 1 THEN 235
00250 : PRINT "FREQUENCY PLAYER B SHOWS 1 FINGER "
00255 : INPUT "(NUMBER BETWEEN 0 AND 1!)"; PB
00260 : IF PB < 0 OR PB > 1 THEN 255
00270 :
00280 : LET A(1,1) = -2 : LET A(1,2) = 3 : REM PAYMENT TABLE
00290 : LET A(2,1) =  3 : LET A(2,2) = -4
00300 :
00310 : LET ZZ = RND(-TI) : REM INITIALIZATION OF RANDOM
00315 :                     REM GENERATOR
00320 :
00330 REM === CONTINUING THE GAME =========================
00340 :
00350 : LET A = 0 : REM PAYMENT TO PLAYER A
00360 :
00370 : FOR I = 1 TO N
00380 :
00390 :    LET ZA = RND(1) : LET ZB = RND(1)
00400 :
00410 :    IF ZA > PA THEN 480
00420 :
00430 :    REM --- PLAYER A SHOWS 1 FINGER
00440 :
00450 :       IF ZB <= PB THEN LET A = A + A(1,1) : GOTO 530
00460 :                        LET A = A + A(1,2) : GOTO 530
00470 :
00480 :    REM --- PLAYER A SHOWS 2 FINGERS
00490 :
00500 :       IF ZB <= PB THEN LET A = A + A(2,1) : GOTO 530
00510 :                        LET A = A + A(2,2)
00520 :
00530 : NEXT I
00540 :
00550 REM === SCORE ===================================
00560 :
00570 : PRINT
00580 : LET M = A/N                      :REM AVERAGE PAYMENT
```

```
00590 : LET M = INT(M*100 + 0.5)/100          :REM ROUNDS-OFF M
00600 : PRINT "AVERAGE PAYMENT FOR PLAYER A PER ROUND: "; M
00610 :
00620 : END
```

Let's confirm these results with data. Assume that Alex throws 1 finger 3/4 of the time and 2 fingers 1/4 of the time (in random sequence). Then Beth can counter this with the simple strategy of always throwing 1 finger, as shown by this tree diagram:

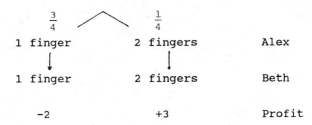

Playing this way Beth wins in the long run because
$$-2 * (3/4) + 3 * (1/4) = -(3/4)$$
tells us that she wins 3/4 of a penny per round. The simulation we tried showed that Alex lost 0.6 penny per round. This tells us that the number of 1's in Alex's strategy is too high.

Assuming Alex chooses the strategy of 1 finger 50% of the time, then Beth can always fall back on the strategy of always throwing 2 fingers, as this tree diagram shows:

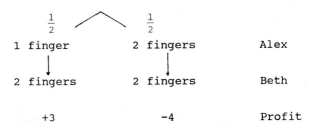

Playing this way Beth wins 1/2 penny per round. Here is the math to show it:
$$3 * (1/2) - 4 * (1/2) = -1 * (1/2)$$
(Compare this with the value in the simulation!)

We can conclude from this that if there exists an optimal frequency of 1's for Alex, then it lies between 0.5 and 0.75. How do we find this number? Let P_A = the probability for Player A's throwing 1 finger, and let P_B = the probability for Player B throwing 1 finger. Then the tree diagram tells us:

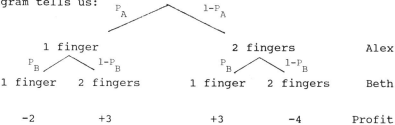

| 1 finger | | 2 fingers | | Alex |

| 1 finger | 2 fingers | 1 finger | 2 fingers | Beth |

| -2 | +3 | +3 | -4 | Profit |

This tells us the average profit per round:

$$w = (-5P_A + 3) * P_B + (7P_A - 4) * (1-P_B)$$

Alex wants to assure himself of an average profit independent of Beth's game plan. This number must apply to random values of P_B, that is, $P_B = 1$, and $P_B = 0$. It follows therefore:

$$w = -5P_A + 3 \text{ and } w = 7P_A - 4; \text{ so } -5P_A + 3 = 7P_A$$

which gives the result of $P_A = 7/12 \approx 58\%$. This means that if Alex chooses 1 finger 7/12 of the time, then in the long run he will make a certain profit -- independent of Beth's strategy. But is this profit a negative or a positive one? Or, to put it another way, is he going to win or lose if he plays this way? The question is easily answered. All we have to do is insert the probability factor we have found into the equation above:

$$w = (-5*(7/12) + 3) * P_B + (7*(7/12) - 4) * (1 - P_B)$$
$$= (1/12) * P_B + (1/12) * (1 - P_B)$$
$$= (1/12)$$

This show us that Alex's profit (or, the guaranteed minimum win in the long run) is positive. The strategy says: show 1 finger 7/12 of the time, and show 2 fingers 5/12 of the time, and Alex will win 1/12 penny per round. So you see, the game isn't exactly fair.

Beth's optimal strategy can be calculated the same

way. She shows 1 finger 7/12 of the time and 2 fingers 5/12
of the time and by doing so she suffers an average loss of
1/12 penny per round. This makes it clear how we have to
write our program.

```
00110 : PRINT CHR$(147)
00120 : PRINT "              PENNY TOSS
00130 : PRINT "              ----------
00140 :
00150 REM GAME USING MIXED STRATEGY
00160 :
00170 REM === RULES OF GAME AND BEGINNING VALUES ===========
00180 :
00190 : PRINT
00200 : PRINT "RULES:
00210 : PRINT " YOU AND THE COMPUTER WILL EACH SHOW 1 OR 2
00220 : PRINT " FINGERS. IF THE TOTAL NUMBER OF FINGERS
00230 : PRINT " SHOWING IS EVEN, I'LL PAY YOU THAT AMOUNT.
00235 : PRINT " IF THE NUMBER OF FINGERS IS ODD,
00240 : PRINT " YOU PAY ME THAT AMOUNT. " : PRINT
00250 :
00260 : LET KA = 100 : LET KB = KA : REM PLAYER'S BANK
00270 :
00280 : PRINT "WE WILL EACH START A BANK IN THE AMOUNT
00290 : PRINT "OF ";KA;"DOLLARS." : PRINT : PRINT
00300 : PRINT "HOW MANY ROUNDS SHALL WE PLAY";
00310 : INPUT N : PRINT
00320 :
00330 : LET G = INT(KA + N/5 + 30/N) : REM LIMIT ON
00335 :                                 REM WINNINGS
00340 :
00350 : PRINT "IN ORDER TO WIN YOU NEED ";G;" DOLLARS
00360 :
00370 : PRINT : PRINT : PRINT "SHALL I PLAY ROUGH ? (Y/N)
00380 :
00390 : GET A$ : IF A$ = "" THEN 390
00400 : IF A$ = "Y" THEN LET PA = 7/12 : GOTO 430
00410 :                LET PA = 5/12
00420 :
00430 : PRINT CHR$(147) CHR$(17) TAB(26) "ME"; TAB(34) "YOU
00440 : PRINT TAB(25) "_____
00450 :
00460 : LET ZZ = RND(-TI) : REM RANDOMIZATION
00470 :
00480 REM === GAME =====================================
00490 :
00500 : FOR I = 1 TO N
00510 :
00520 :    PRINT : PRINT I". ROUND:
00530 :
00540 :    REM --- COMPUTER'S MOVE
```
240

```
00550 :
00560 :     LET Z = RND(1)
00570 :     IF Z < PA THEN LET F = 1 : GOTO 600
00580 :                     LET F = 2
00590 :
00600 :  REM --- PLAYER'S MOVE
00610 :
00620 :     INPUT "YOUR CHOICE (1 OR 2) ";B$
00630 :     IF B$ <> "1" AND B$ <> "2" THEN 620
00640 :     LET B = VAL(B$)
00650 :
00660 :  REM --- SCORE
00670 :
00680 :     PRINT "MY CHOICE WAS: "; F
00690 :
00700 :     IF F <> B THEN KA = KA+3 : KB = KB-3 : GOTO 760
00710 :     IF F = 1  THEN KA = KA-2 : KB = KB+2 : GOTO 750
00720 :                    KA = KA-4 : KB = KB+4
00730 :
00740 :     PRINT : PRINT "PAYMENT TO YOU: 4"; : GOTO 780
00750 :     PRINT : PRINT "PAYMENT TO YOU: 2"; : GOTO 780
00760 :     PRINT : PRINT "PAYMENT TO ME:  3";
00770 :
00780 :     PRINT TAB(26) KA; TAB(34) KB
00790 :
00800 :  REM --- TO GAME EVEN?
00810 :
00820 :     IF KB < G THEN 860 : REM THE GAME CONTINUES
00830 :
00840 :     PRINT : PRINT CHR$(18) "YOU HAVE WON!": GOTO 880
00850 :
00860 : NEXT I
00870 :
00880 : END
```

****

It's time for another literary digression. Edgar Allan
Poe's story "The Purloined Letter" (1845) contains a wonderful passage that shows how the use of various mixed strategies can master a situation. Here, Poe's great detective
Dupin tells his friend a story to demonstrate a point of
deduction.

"I knew a [school-boy] about eight years of age, whose success at guessing in the game of 'even and odd' attracted
universal admiration. This game is simple and is played
with marbles. One player holds in his hand a number of

these toys, and demands of another whether that number is even or odd. If the guess is right, the guesser wins one; if wrong, he loses one. The boy to whom I allude won all the marbles of the school. Of course he had some principle of guessing; and this lay in mere observation and measurement of the astuteness of his opponents. For example, an errant simpleton is his opponent, and, holding up his closed hand, asks, 'are they even or odd'? Our schoolboy replies, 'odd,' and loses; but upon the second trial he wins, for then he says to himself, 'the simpleton had them even upon the first trial, and his amount of cunning is just sufficient to make him have them odd upon the second; I will therefore guess odd;' -- he guesses and wins. Now, with a simpleton a degree above the first, he would have reasoned thus: 'This fellow finds that in the first instance I guessed odd, and in the second, he will propose to himself upon the first impulse, a simple variation from even to odd, as did the first simpleton; but then a second thought will suggest that this is too simple a variation, and finally he will decide upon putting it even as before. I will therefore guess even;' -- he guesses even, and wins. Now this mode of reasoning in the schoolboy, whom his fellows termed 'lucky,'-- what, in the last analysis, is it?"

"It is merely," I said, "an identification of the reasoner's intellect with that of his opponent."

"It is," said Dupin.

****

BRAINTEASERS

1. This time Alex and Beth are going to play a bit differently. Each one throws either 2 or 3 fingers. If the total is even, Alex pays; if the number is odd, Beth pays. Try this out on the computer and see who has the advantage in this game. Develop the optimal strategies for the players.

2. Players A and B each lay a coin on the table and cover it. Then they uncover their coins. If both coins show HEADs, player A pays B 1.5 times the unit of currency they are playing with. (You'll have to use dollars or, if you are playing with pennies, keep a point score until the end.) If both coins show TAILs, A pays B 0.5 units of currency. If the coins turn up one HEAD and the other TAIL, then B pays A one unit of currency. The computer, as player A, has to be able to protect itself with a certain measure of skill.

3. In Italy thay play a form of choosing up called <u>Morra</u>. When the players shoot one, two, or three fingers, they simultaneously call out the predicted total of both players' fingers.

Run a simulation of this game to show that you can never lose in the long run if you always guess 4 and also show 1 finger 5 times out of 12; 2 fingers 4 times out of 12; and 3 fingers 3 times out of 12.

4. Player A has a playing card with figures on each side. On one side is a RED ACE, and on the other is a BLACK 5. Player B has a card with a BLACK 2 and a RED 4 on either side. Each player chooses one side of his card and then both players show their cards simultaneously. If the colors match, then A is the winner, if not, B wins. Each player wins from the other the amount that the opponent's card is worth.

Question: does one player have an advantage and, if so, who? A computer program that simulates the game can help you find the answer.

5. Write a program for the game Odd-Even and try to incorporate something like the strategy Poe described for his school-boy.

****

Example 7.2   GUESS THE CARD

We need 11 playing cards with values from Ace to Jack. The Jack is worth 11 points. The cards are shuffled, one is drawn at random and laid face down on the table. The players do not know the value of this card. The remaining 10 cards are divided between the players so that each has 5 cards. The object is to guess the unknown card. To do this, players ask each other questions such as, "Do you have such-and-such a card?" Players have to answer with the truth about the cards they hold. Instead of asking the question, a player can break off the game at any time by risking a "prediction." This consists of "calling" the card on the table, or trying to guess it. If this player's call is correct, he wins the game. If his call is wrong, he loses.

Here is an explanation of the game based on Martin Gardner's description.

In order to play this game well, every player has to try to get as much information as possible and to divulge as little as possible until he thinks he knows enough to be able to risk a "call". The appeal of the game lies in bluffing, ie., asking for a card in one's own hand. If a player never bluffed, the opponent would know immediately when asked for a card not in his hand, that the card must be face down on the table. Thus, he would "call" and win. Bluffing is a crucial defensive as well as offensive strategy that can persuade the opponent to make a false call.

When player A asks for a card -- let's say, for the jack -- and the answer is "yes", then both players know that B has this card. It is therefore removed from play. If B doesn't have the jack, he answers "no". If he thinks A is not bluffing, he calls the jack and wins if his suspicion was right. If he does not call and the covered card is the jack, then A will call it in the next round. So, if A then does not call the jack in the next round, it just means that he has it in his hand and was bluffing when he asked for it. Because both players know this, this card is also then re-

moved from play. This is how each player's hand gets smaller with successive rounds. The program we need to write has
to make possible a dialogue like this:

```
Guess the Card

You are dealt the cards: 5 7 2 1 4.
Do you want to begin?  (y/n) y
Care to risk a call?  (y/n) n
  Which card? 3
I hold that one.
Do you hold the 8? (y/n) n
Care to risk a call?  (y/n) n
  Which card? 6
I hold that one.
Do you hold the 10? n
Care to risk a call?  (y/n) n
  Which card? 2
I don't hold that one.
I think the card on the table is the 11.
The card on the table is the 9.
Congratulations, you win.
Another round?  (y/n) n
Goodbye.
```

The strategy that the computer uses can be described as
follows. It involves these possibilities:

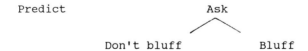

                Predict   Ask

            Don't bluff   Bluff

(1) The computer will risk a call under the following circumstances:

(a) If it knows the card on the table. This happens
when the opponent has no more unknown cards, or if the computer -- without bluffing -- has asked about a card and
gotten "no" for an answer and then the opponent does not

risk a prediction about the card on the table. In this case, the computer then calls the card.

(b) If the computer has no cards but the player has one or more. (The computer knows that in the next round the player, knowing the card on the table, will risk a call.)

(c) If the computer has answered a question about a card with "no." In this case it decides whether to risk a call by spinning a roulette wheel that is divided into probability segments p( ). (Unfortunately we can't go into how the individual segments were calculated.)

The card the computer guesses in the cases of (b) and (c) is determined by chance.

(2) If the computer does not call the card, then it can ask about the player's cards. The decision to bluff or not is again determined by chance.

```
00100 : PRINT CHR$(147)
00110 : PRINT "          GUESS THE CARD
00120 : PRINT "          --------------
00130 :
00140 REM BASED ON A GAME BY RUFUS ISAAKSOHN
00150 :
00160 REM ===================== DIRECTIONS ================
00170 :
00180 : PRINT
00190 : PRINT "WOULD YOU LIKE TO SEE THE RULES?(Y/N) ";
00200 : GET A$ : IF A$ = "" THEN 200
00210 : PRINT A$
00220 : IF A$ <> "Y" THEN 260
00230 :
00240 : REM PUT THE RULES OF THE GAME HERE.
00250 :
00260 REM ==================== INITIALIZATION =============
00270 :
00280 : LET SP = 0 : REM NUMBER OF ROUNDS PLAYED
00290 : LET VE = 0 : REM NUMBER OF ROUNDS LOST BY COMPUTER
00300 :
00310 : DIM P(7,7) : REM TABLE OF PROBABILITIES
00320 : DIM M(11)  : REM SHUFFLING THE CARDS
00330 : DIM C(5)   : REM TABLE OF COMPUTER'S CARDS
00340 : DIM U(6)   : REM TABLE OF UNKNOWN CARDS
00350 : DIM A(11)  : REM TABLE OF TURNED OVER CARDS
00360 :
00370 REM ==================== PROBABILITIES ==============
00380 :
00390 : LET P(0,0) = 0
```

```
00400 : FOR I = 1 TO 5
00410 :   FOR J = 1 TO 5
00415 :     LET T = P(J,I-1)
00420 :     LET P(I,J) = (1+J*T*(1-P(J-1,I))) / (1+(J+2)*T)
00425 :     LET T = P(I,J-1)
00430 :     LET P(J,I) = (1+I*T*(1-P(I-1,J))) / (1+(I+1)*T)
00440 :   NEXT J
00450 : NEXT I
00460 :
00470 REM ======================= ROUND ===================
00480 :
00490 : REM --- BEGINNING VALUES
00500 :
00510 :   LET SP = SP + 1      : REM NUMBER OF ROUNDS
00520 :   LET E$ = "CONTINUE" : REM FLAG THAT DETERMINES
00525 :                         REM WHEN GAME ENDS
00530 :   LET M$ = "UNKNOWN"  : REM COVERED CARD (STILL)
00535 :                         REM UNKNOWN
00540 :
00550 :   FOR I = 1 TO 11 : LET A(I) = 0 : NEXT I
00560 :
00570 :   LET CK = 5 : REM NUMBER OF COMPUTER'S CARDS
00580 :   LET UK = 6 : REM NUMBER OF UNKNOWN CARDS
00590 :   LET AK = 0 : REM NUMBER OF UNCOVERED CARDS
00600 :
00610 : REM --- SHUFFLE
00620 :
00630 :   FOR I = 1 TO 11 : LET M(I) = I : NEXT I
00640 :
00650 :   FOR I = 1 TO 11
00660 :     LET R = INT(RND(1)*(12-I)) + I
00670 :     LET V = M(R) : LET M(R) = M(I) : LET M(I) = V
00680 :   NEXT I
00690 :
00700 : REM --- DEALING THE CARDS
00710 :
00720 :   FOR I = 1 TO 5  : LET C(I) = M(I)    : NEXT I
00730 :
00740 :   FOR I = 6 TO 11 : LET U(I-5) = M(I) : NEXT I
00750 :
00760 :   LET VK = U(6)   : REM COVERED CARD
00770 :
00775 :   PRINT CHR$(147) : PRINT : PRINT : PRINT
00780 :   PRINT "YOU HAVE THESE CARDS: ";
00790 :   FOR I = 1 TO 5 : PRINT U(I);"   ";: NEXT I
00800 :
00810 : REM --- DETERMINING WHO PLAYS NEXT
00820 :
00830 :   LET DR$ = "COMPUTER"
00840 :
00850 :   PRINT
00860 :   PRINT : PRINT "[]DO YOU WANT TO BEGIN?(Y,N) ";
```

```
00870 :    GET A$ : IF A$ = "" THEN 870
00880 :    PRINT A$
00890 :    IF A$ = "Y" THEN LET DR$ = "PLAYER"
00900 :
00910 : REM --- QUESTION AND ANSWER
00920 :
00930 :    IF DR$ = "PLAYER" THEN 960 : REM PLAYER BEGINS
00940 :    GOSUB 2000            : REM COMPUTER'S QUESTIONS
00950 :    IF E$ = "END" THEN 990      : REM UNCOVER CARDS
00960 :    GOSUB 2610            : REM PLAYER'S QUESTIONS
00970 :    IF E$ <> "END" THEN 940  : REM ROUND CONTINUES
00980 :
00990 : REM --- UNCOVER CARDS
01000 :
01005 :    PRINT : PRINT
01010 :    PRINT "THE COVERED CARD IS THE"; VK; ". ";
01020 :
01030 :    IF GW$ = "COMPUTER" THEN PRINT "I WIN.": GOTO 1070
01040 :    PRINT "CONGRATULATIONS! YOU WIN."
01050 :    LET VE = VE + 1
01060 :
01070 : REM --- REPEAT?
01080 :
01090 :    PRINT : PRINT : PRINT "ANOTHER ROUND? (Y/N)";
01100 :    GET A$ : IF A$ = "" THEN 1100
01110 :    PRINT A$
01120 :    IF A$ = "Y" THEN 470
01130 :
01140 : REM --- EXIT
01150 :
01160 :    PRINT : PRINT : PRINT "WE PLAYED"; SP;" ROUNDS,
01170 :    PRINT "YOU WON";VE;" OF THEM ";
01180 :    PRINT "AND LOST"; SP - VE;".
01190 :
01200 :    IF VE < SP/2 THEN PRINT : PRINT "IT'S BEEN FUN! ";
01210 :    PRINT "SO LONG!"
01220 :
01230 :    END
01240 :
01250 REM === END OF ROUND =================================
01260 :
01270 :
02000 REM +++ SUBROUTINE 'COMPUTER'S QUESTIONS' +++++++++++
02010 :
02020 : REM --- DECISION TO ANNOUNCE OR ASK
02030 :
02040 :    IF M$ = "KNOWN" THEN 2090 : REM COVERED CARD IS
02045 :                              REM KNOWN
02050 :    IF CK = 0 OR UK = 1 OR CK + UK = 3 THEN 2160
02055 :                              REM CALL
02060 :    IF RND(1) < 1/(1 + UK * P(UK-1,CK-1)) THEN 2240
02065 :                              REM BLUFFING
```

248

```
02070 :    GOTO 2400 :                    REM NO BLUFFING
02080 :
02090 : REM --- COMPUTER CALLS CARD IT HAD ASKED FOR BEFORE
02100 :
02110 :    LET T = A(L) : REM LAST UNCOVERED CARD IS CALLED
02120 :    GOSUB 5000   : REM COMPUTER CALLS
02130 :
02140 :    RETURN
02150 :
02160 : REM --- COMPUTER CALLS THE RANDOMLY CHOSEN CARD
02170 :
02180 :    LET R = INT(6*RND(1))+1 : IF U(I) = 0 THEN 2180
02190 :    LET T = R : REM RANDOM CARD IS CALLED
02200 :    GOSUB 5000   : REM COMPUTER CALLS
02210 :
02220 :    RETURN
02230 :
02240 : REM --- COMPUTER BLUFFS
02250 :
02260 :    LET R = INT(5*RND(1))+1 : IF C(R) = 0 THEN 2260
02270 :
02280 :    PRINT "DO YOU HAVE THE";C(R);"?(Y/N) ";
02290 :    GET A$ : IF A$ = "" THEN 2290
02300 :    PRINT A$
02310 :
02320 :    LET AK = AK + 1  : REM ONE ADDITIONAL UNCOVERED
02325 :                        REM CARD
02330 :    LET A(AK) = C(R) : REM ENTER THE UNCOVERED CARD
02340 :    LET C(R) = 0     : REM ERASE UNCOVERED CARD
02350 :    LET L = AK       : REM LAST MOVE
02360 :    LET CK = CK - 1  : REM COMPUTER HAS ONE LESS CARD
02370 :
02380 :    RETURN
02390 :
02400 : REM --- COMPUTER DOES NOT BLUFF
02410 :
02420 :    LET R = INT(6*RND(1))+1 : IF U(R) = 0 THEN 2420
02430 :
02440 :    PRINT "DO YOU HAVE THE"; U(R);"?(Y,N) ";
02450 :    GET A$ : IF A$ = "" THEN 2450
02460 :    PRINT A$
02470 :    IF A$ = "Y" THEN 2490
02475 :    IF A$ <>"N" THEN PRINT "'Y' OR 'N'!" : GOTO 2440
02480 :
02490 :    LET AK = AK + 1  : REM ONE ADDITIONAL UNCOVERED
02495 :                        REM CARD
02500 :    LET A(AK) = U(R) : REM ENTER THE UNCOVERED CARD
02510 :    LET U(R) = 0     : REM ERASE THE UNCOVERED CARD
02520 :    LET L = AK       : REM LAST MOVE
02530 :    LET UK = UK - 1  : REM ONE UNCOVERED CARD LESS
02540 :
02550 :    IF A$ = "N" THEN M$ = "KNOWN" : REM THE COVERED
```

```
02555 :                                    REM CARD IS KNOWN
02560 :
02570 :    RETURN
02580 :
02590 REM END OF COMPUTER'S QUESTIONS ++++++++++++++++++++++
02600 :
02610 REM +++ SUBROUTINE PLAYER'S QUESTIONS ++++++++++++++++
02620 :
02630 : REM --- START THE QUESTIONING
02640 :
02650 :    PRINT "DO YOU WANT TO TRY A CALL?(Y/N) ";
02660 :    GET A$ : IF A$ = "" THEN 2660
02670 :    PRINT A$
02680 :    IF A$ = "Y" THEN 4000 : REM PLAYER CALLS
02690 :
02700 :    INPUT "WHICH CARD "; K
02710 :    IF K < 1 OR K > 11 THEN 2700
02720 :
02730 : REM --- TEST, TO SEE IF THE CARD HAS ALREADY BEEN
02735 : REM      ASKED FOR
02740 :
02750 :    LET J = 0
02760 :    FOR I = 1 TO AK
02770 :       IF A(I) = K THEN LET J = I
02780 :    NEXT I
02790 :    IF J = 0 THEN 2810
02795 :    PRINT "THAT'S ALREADY BEEN ASKED FOR!": GOTO 2700
02800 :
02810 : REM --- TEST, TO SEE IF CARD IS IN COMPUTER'S HAND
02820 :
02830 :    LET J = 0
02840 :    FOR I = 1 TO 5
02850 :       IF C(I) = K THEN LET J = I
02860 :    NEXT I
02870 :    IF J = 0 THEN 2980 : REM COMPUTER DOES NOT HAVE
02875 :                          REM CARD
02880 :
02890 :    PRINT "              ... I HAVE IT.
02900 :
02910 :    LET AK = AK + 1  : REM ONE ADDITIONAL UNCOVERED
02915 :                       REM CARD
02920 :    LET A(AK) = C(J) : REM ENTER THE UNCOVERED CARD
02930 :    LET C(J) = 0     : REM ERASE THE UNCOVERED CARD
02940 :    LET CK = CK - 1  : REM COMPUTER HAS ONE LESS CARD
02950 :
02960 :    RETURN
02970 :
02980 :    PRINT "              ... I DON'T HAVE THAT ONE!
02990 :
03000 : REM --- REACTION TO PLAYER'S QUESTIONS
03010 :
03020 :    IF CK = 0 OR UK = 1 THEN 3290 : REM COMPUTER CALLS
```

```
03030 :
03040 :    LET Z = (CK+1) * P(CK, UK-2) - CK * P(CK-1,UK-1)
03050 :    LET N = 1 + (CK+1) * P(CK, UK-2)
03060 :    IF RND(1) < Z/N THEN 3220      : REM COMPUTER CALLS
03070 :
03080 :    IF UK = 2 THEN 3290  : REM COMPUTER CALLS RANDOM
03085 :                              REM CARD
03090 :
03100 :    LET J = 0
03110 :    FOR I = 1 TO 6
03120 :      IF U(I) = K THEN LET J = I
03130 :    NEXT I
03140 :
03150 :    LET AK = AK + 1  : REM ONE ADDITIONAL UNCOVERED
03155 :                         REM CARD
03160 :    LET A(AK) = U(J) : REM ENTER THE UNCOVERED CARD
03170 :    LET U(J) = 0     : REM ERASE UNCOVERED CARD
03180 :    LET UK = UK - 1  : REM ONE LESS UNKNOWN CARD
03190 :
03200 :    RETURN
03210 :
03220 : REM --- COMPUTER CALLS PLAYER'S LAST MOVE
03230 :
03240 :    LET T = K  : REM LAST CARD NAMED BY PLAYER IS
03245 :                    REM CALLED
03250 :    GOSUB 5000 : REM COMPUTER CALLS
03260 :
03270 :    RETURN
03280 :
03290 : REM --- COMPUTER CALLS PLAYER'S LAST MOVE
03300 :
03310 :    LET R = INT(6*RND(1))+1 : IF U(R) = 0 THEN 3310
03320 :    LET T = R : REM RANDOM CARD IS CALLED
03330 :    GOSUB 5000: REM COMPUTER CALLS
03340 :
03350 :    RETURN
03360 :
04000 REM +++ SUBROUTINE 'PLAYER CALLS' +++++++++++++++++++++
04010 :
04020 : PRINT : PRINT "WHAT CARD ARE YOU ANNOUNCING";
04030 : INPUT A
04040 :
04050 : LET GW$ = "COMPUTER"              : REM COMPUTER WINS
04060 : IF A = VK THEN LET GW$ = "PLAYER" : REM PLAYER WINS
04070 : LET E$ = "END"
04080 :
04090 : RETURN
04100 :
04110 REM END OF SUBROUTINE "PLAYER CALLS' +++++++++++++++++
04120 :
05000 REM +++ SUBROUTINE 'COMPUTER CALLS' +++++++++++++++++
05010 :
```

```
05020 : PRINT "I THINK THE COVERED CARD IS THE "; T
05030 :
05040 : LET GW$ = "PLAYER"                : REM PLAYER WINS
05050 : IF T = VK THEN GW$ = "COMPUTER" : REM COMPUTER WINS
05060 : LET E$ = "END"
05070 :
05080 : RETURN
05090 :
05100 REM END OF SUBROUTINE 'COMPUTER CALLS' +++++++++++++++
05110 :
```

****

BRAINTEASERS

6. Build a security measure into the game "Guess the Card"
to prevent a player from answering a question about his own
cards with anything but the truth.

7. Player A tosses a coin but hides it on the table from his
opponent. If it is HEADS he can call it as such and be paid
a dime by player B.   If it is TAILS, player A can bluff,
call HEADS, and get his dime. Or, he can announce TAILS and
pay player B the dime. Player B can believe A and pay the
dime or he can demand proof.   In this case A uncovers the
coin. If it is HEADS, B pays 2 dimes, otherwise A pays 2
dimes to B. Write a program for this with the computer as
player A.

CHAPTER VIII

PUZZLES

You could call the activities in this chapter Robinson
Crusoe games, or games of concentration, as easily as "puzzles." They are thought games, mental gymnastics for the
solitary player -- in other words, true brainteasers. But
instead of being stranded on a desert island, consider yourself alone with your video screen. Novices are advised to
store in provisions: you may be at it a while.

Let's start with something that has enthralled the
intelligence of speculative thinkers for ages: magic
squares. The origin of this haunting configuration of numbers (or sometimes letters) is shrouded in the mists of
time. Supposedly, one can trace back to ancient China this
harmonious ordering of integers in a quadrant in which the
sum of every column, row, and diagonal is identical. The
simplest and oldest magic square looks like this:

$$4 \quad 9 \quad 2$$
$$3 \quad 5 \quad 7$$
$$8 \quad 1 \quad 6$$

According to an old legend this square came to the legendary
emperor Yu of the Shang dynasty (c. 2000 B.C.) on the bank
of the river Lo in China. Hence the square was called the
"Lo-Shu", which means Lo-Document. For the Chinese the even
numbers represent Yin, the female principle; the odd numbers
represent Yang, the male principle. The Lo-Shu embodies the
balance of these forces. It is therefore the symbol of cosmic harmony.

Magic squares were important to the astrologers in the
16th and 17th centuries when they were used in conjunction
with the other lore surrounding the pre-Copernican world
view. In this primitive cosmography the seven known plan-
ets, or heavenly bodies (the earth's moon, Mercury, Venus,
the sun, Mars, Jupiter, and Saturn) were thought to revolve
around the earth. Saturn, the remotest (and therefore con-

sidered the tiniest planet) had the smallest possible magic
square associated with it:

```
        2   9   4
        7   5   3
        6   1   8
```

The astrologer Paracelsus (1493-1541), wrote about the magi-
cal properties of an alchemical amulet, a seal depicting a
magic square that was supposed to emody the powers of the
planet Saturn.

    "This Saturn seal is to made of fine, pure, unadult-
erated lead from Villach. On one side appears the planet's
square consisting of three rows each adding up to fifteen.
On the other side shall be the picture of the planet: an old
man with a long beard digging in the earth with a gravedig-
ger's shovel. On his head is a star and the name Saturnus.
    First, this seal is very good for a pregnant woman. If
she carries it with her, nothing will go wrong in child-
birth. Secondly, no matter where one places this seal, it
will grow and increase. If a horseman carries this seal
with him in his left boot, nothing ill shall befall his
horse.
    If, however, this seal is made when Saturn is on the
wane, on a Saturday and in the hour of the planet, then it
will prevent all good, and all that it touches will daily
decrease til dissolution. And should this seal then be se-
creted among soldiers, they will never enjoy fortune, but
soon break up and disband."

Who would want to vouch for these extravagant properties of magic squares nowadays?   Nonetheless, their construction presents one of the oldest mathematical problems
known. Let's give the computer this task.

Example 8.1 MAGIC SQUARES
Our object is to develop a technique for constructing and
programming magic squares.

We begin with an odd number, N, of rows and columns. First,
let's order the numbers $1,2,3,\ldots N^2$ into a square format.
The sums of the rows, columns and diagonals must be constant, i.e., the same. These so-called magic constants are
calculated like this:

$$M = \frac{1}{N} * (1 + 2 + \ldots + N^2) = \frac{1}{N} * \frac{N^2(N^2 + 1)}{2} = \frac{N(N^2 + 1)}{2}$$

For N = 3 we have M = 15; but for N = 5, we have M = 65.
The easiest way to derive an algorithm is to analyze a familiar magic square and figure out the priciples behind it.
Let's try that with the Lo-shu square. The principle behind
its construction becomes more evident when we expand it into
a larger checkerboard:

```
4 9 2 4 9 2 4 9 2 4 9 2 4 9 2 4 9 2
3 5 7 3 5 7 3 5 7 3 5 7 3 5 7 3 5 7
8 1 6 8 1 6 8 1 6 8 1 6 8 1 6 8 1 6
4 9 2 4 9 2 4 9 2 4 9 2 4 9 2 4 9 2
3 5 7 3 5 7 3 5 7 3 5 7 3 5 7 3 5 7
8 1 6 8 1 6 8 1 6 8 1 6 8 1 6 8 1 6
4 9 2 4 9 2 4 9 2 4 9 2 4 9 2 4 9 2
3 5 7 3 5 7 3 5 7 3 5 7 3 5 7 3 5 7
8 1 6 8 1 6 8 1 6 8 1 6 8 1 6 8 1 6
4 9 2 4 9 2 4 9 2 4 9 2 4 9 2 4 9 2
3 5 7 3 5 7 3 5 7 3 5 7 3 5 7 3 5 7
8 1 6 8 1 6 8 1 6 8 1 6 8 1 6 8 1 6
4 9 2 4 9 2 4 9 2 4 9 2 4 9 2 4 9 2
3 5 7 3 5 7 3 5 7 3 5 7 3 5 7 3 5 7
8 1 6 8 1 6 8 1 6 8 1 6 8 1 6 8 1 6
```

Right away you notice the path traced through the numbers by the arrows. We have to construct an analogous path to be able to create a magic square of the fifth order.

The next step is deriving a square from this key. As with Lo-shu, we begin with the 1 in the middle field (square) of the bottom-most row (row 5, column 3).

			2	
				3
4	6			
	5	7		
		1	8	

(with a "4" to the right of row 3, and a "2" below row 5, column 3)

To prevent the 2 from lying outside the square, we have to imagine this square having a three-dimensional structure. In other words, imagine that the bottom and top margins of the square are glued together to make a cylinder. This automatically places the 2 in row 1, column 4. The 3 then goes in row 2, column 5, according to the arrow path we drew

256

above. To prevent the 4 from lying outside the square, we
imagine the same three-dimensional structure applying to the
sides of the square. Picture the right and left margins
glued together. The result resembles something like a bicy-
cle inner tube -- a structure mathematicians call a "torus."
Now the 4 goes in the square defined as row 3, column 1.
Continue in the same way. After placing every Nth number,
we come to an occupied square. This is the spot where we
make a short jog upwards in our pattern (see the arrow
path). The row number decreases by 1, and the column stays
the same. The square constructed on this principle looks
like this:

11	18	25	2	9
10	12	19	21	3
4	6	13	20	22
23	5	7	14	16
17	24	1	8	15

Note that the element $(1/2) * (N^2 + 1) = 13$ is in the center. The sum of any two numbers on opposite sides and equi-
distant from this center element is $N^2 + 1 = 26$.

This method of constructing magic squares was brought
back from Siam by a French ambassador named S. de la Loubere
in 1687. Therefore, it is generally called the "Siamese method." It's general rules can be summarized as follows:

SIAMESE METHOD OF CONSTRUCTING MAGIC SQUARES OF ODD ORDER

1. Write 1 in the center square of the Nth row.
2. If there is a number in square (i,j), then the next
   number goes in square (i+1, j+1), if this square is
   not already occupied. Otherwise, it goes to square
   (i-1, j).
3. If an index exceeds the order N, then it is replaced
   by 1; if it is less than 1, it is replaced by N.

The following program as constructed according to this algorithm.

```
00100 : PRINT CHR$(147)
00110 : PRINT "          MAGIC SQUARES
00120 : PRINT "          -------------
00130 :
00140 REM GENERATES MAGIC SQUARES OF ODD ORDER
00150 REM ACCORDING TO THE SIAMESE METHOD (DE LA LOUBERE)
00160 :
00170 REM === INPUT ========================================
00180 :
00190 : PRINT
00200 : INPUT "ORDER (ODD NUMBER) "; N
00210 : IF N < 3 OR 2*INT(N/2) = N THEN 200
00220 :
00230 REM === INITIALIZATION ===============================
00240 :
00250 : DIM Q(N,N)
00260 :
00270 : LET I = N-1 : LET J = (N-1)/2 : REM PRE-POSITION OF
00275 :                                 REM "1"
00280 :
00290 REM === ENTERING THE NUMBERS =========================
00300 :
00310 : FOR U = 0 TO N-1
00320 :    FOR V = 0 TO N-1
00330 :       IF I < N THEN LET I = I+1 : GOTO 350
00340 :                   LET I = 1
00350 :       IF J < N THEN LET J = J+1 : GOTO 370
00360 :                   LET J = 1
00370 :       LET Q(I,J) = U*N+V+1 : REM ENTERING THE NUMBER
00380 :    NEXT V
00390 :    LET I = I-2 : LET J = J-1   : REM JOG UPWARDS
00400 : NEXT U
00410 :
00420 REM === DISPLAY ======================================
00430 :
00440 : PRINT : PRINT
00450 : FOR I = 1 TO N
00460 :    LET S = 0
00470 :    FOR J = 1 TO N
00480 :       PRINT TAB(5*J) Q(I,J);
00490 :       LET S = S + Q(I,J)
00500 :    NEXT J
00510 :    PRINT TAB(5*(N+1)) "----"; S
00520 :    PRINT
00530 : NEXT I
00540 :
00550 : FOR J = 1 TO N : PRINT TAB(5*J) "-----";: NEXT J
00560 :
00570 : PRINT
00580 : FOR J = 1 TO N
00590 :    LET S = 0
00600 :    FOR I = 1 TO N : LET S = S + Q(I,J) : NEXT I
```

258

```
00610 :   PRINT TAB(5*J) S;
00620 : NEXT J
00630 :
00640 : END
```

How do we know that this program produces magic squares of
any odd order?   Mathematicians need proof.   Here is part of
it.

Look again at the arrow path we traced through the
magic square above.   You see it is grouped into five sequences: (1,2,3,4,5),   (6,7,8,9,10),   (11,12,13,14,15),
(16,17,18,19,20), and (21,22,23,24,25).   These can be expressed generally as follows:

sequence u	digit v of the sequence					
	0	1	2	3	...	N-1
0	1	2	3	4	.	N
1	N+1	N+2	N+3	N+4	:	2N
2	2N+1	2N+2	2N+3	2N+4	:	3N
u	....				u*N+v+1	
N-1	....					N^2

Accordingly, every number $z \in \{1,2,3,......,N^2\}$ of the
square can be expressed in the form:

$$z = u*N + v + 1 \quad \text{with} \quad u,v \in \{0,1,2,...,N-1\}$$

In this case u is the number of the sequence and v the
digit in the sequence.   This representation of the mathematical relationships is the one we use in the program in
lines 310-400.

The Siamese method causes every field with sequence
number u and digit v to have the following "neighbors" (adjacent numbers):

u+1, v+1	u+2, v+3
u , v	u+1, v+2

Here the numbers are to be taken modulo N.   For example (N =
5, u = 2, v = 2):

19	21
13	20

→

3*5 + 3 + 1	4*5 + 0 + 1
2*5 + 2 + 1	3*5 + 4 + 1

→

3,3	4,0
2,2	3,4

The zero is derived $2 + 3 = 5 = 0$ (modulo 5). Now we have to show that the sum of row i, column j and both the diagonals always equals M. This is seen from the following: when we add the elements of any one row, every sequence number u and every digit v appears exactly once. Thus, we derive the sum:

$$(0+1+2+...+N-1)*N \quad + \quad (0+1+2+...+N-1) \quad + \quad N*1$$

$$= \frac{N(N-1)}{2} *N + \frac{N(N-1)}{2} + N = \frac{N^3 + N}{2} = M$$

(according to the definition of M)

Analogously, we find the magic constant M as the sum of any column. The element in the middle of the square is $(N^2 + 1)/2$, i.e., the median value of the sequence $1,2,3,...N^2$. In the case where $N = 5$, this is the number 13. Every two elements that are symetrical to the middle element have the sum total $N^2 + 1$. When $N = 5$ this is 26. The sum of each diagonal is

(middle element)

$$\frac{1}{2} * (N-1) * (N^2+1) + \frac{N^2+1}{2} = \frac{N^3+N}{2} = M$$

(number of symetri-
cally lying elements)

Now let's familiarize ourselves with another procedure for generating magic squares that also suggests interesting parallels to the so-called Latin Squares.

First, have another look at our Lo-Shu. As we have just demonstrated, each of its elements can be represented in the form $z = u*3 + v + 1$, with $u,v \in \{0,1,2\}$. When we take the whole square apart we get

4 9 2
3 5 7
8 1 6

= 3 *

1 2 0
0 1 2
2 0 1

+

0 2 1
2 1 0
1 0 2

+

1 1 1
1 1 1
1 1 1

These squares have an extremely interesting structure. The
first square to the right of the equal sign contains three
rows that are derived from each other by a principal of
cyclical exchange. In each row and each column each of the
numbers 0,1,2 appears exactly once. A configuration having
this characteristic is called a Latin Square. We can imagine the next square as derived by turning the first square
90° counter-clockwise. If we superimpose the two, we get:

$$
\begin{array}{ccc}
10 & 22 & 01 \\
02 & 11 & 20 \\
21 & 00 & 12
\end{array}
$$

We can conclude that each of the pairs uv with u,v ϵ
{0,1,2} comes up only once. This kind of square is called
an Euler Square; two Latin Squares that, when superimposed,
form an Euler Square are said to be orthogonal to each other. We can conclude that:

A magic square can always be constructed from orthogonal
Latin Squares.

But how do we get these orthogonal Latin Squares? Here's
some math to help. For an odd order N the following will
work:

MAGIC SQUARES FROM ORTHOGONAL LATIN SQUARES OF AN ODD ORDER

1. The first row is filled with the numbers: 0,1,...,
   n-1, starting anywhere and continuing randomly.
2. The alignment of the succeeding rows is cyclically
   altered: $a_0, a_1, a_2,...a_{n-1}$ becomes $a_{N-1}, a_0, a_1,$
   ..., a_{N-2}. Call this square A.
3. Square A is rotated 90°. The result is B.
4. Q = N*A + B + (1)

The following program applies this algorithm. For example,
it produces:

```
        12    24     1    20     8
         9    11    25     3    17
        16    10    13    22     4
         5    18     7    14    21
        23     2    19     6    15



00100 : PRINT CHR$(147)
00110 : PRINT "          MAGIC SQUARE
00120 : PRINT "          -----------
00130 :
00140 REM CONSTRUCTS MAGIC SQUARES ACCORDING TO THE EULER
00150 REM METHOD. THE SIDE LENGTH MUST BE AN ODD NUMBER
00160 :
00170 REM === INPUT =========================================
00180 :
00190 : PRINT
00200 : INPUT "SIDE LENGTH (ODD NUMBER) "; N
00210 : IF N < 3 OR 2*INT(N/2) = N THEN 200
00220 :
00230 : DIM V(N), A(N,N), Q(N,N)
00240 :
00250 : LET R = RND (-TI) : REM RANDOMIZE
00260 :
00270 REM === CALCULATION ===================================
00280 :
00290 : REM --- SETTING THE RANDOM DIGITS BETWEEN 0 AND N
00300 :
00310 :    FOR J = 1 TO N
00320 :       LET V(J) = INT(N*RND(1))
00330 :       FOR K = 1 TO J-1
00340 :          IF V(J-K) = V(J) THEN 320
00350 :       NEXT K
00360 :    LET A(1,J) = V(J)
00370 :    NEXT J
00380 :
00390 : REM --- CYCLICAL ALTERATION
00400 :
00410 :    FOR I = 2 TO N
00420 :       FOR J = 1 TO N
00430 :          LET L = J - 1 - N*INT((J-2)/N)
00440 :          LET A(I,J) = A(I-1,L)
00450 :       NEXT J
00460 :    NEXT I
00470 :
00480 : REM TURNING AND MULTIPLYING
00490 :
00500 :    FOR I = 1 TO N
00510 :       FOR J = 1 TO N
00520 :          LET Q(I,J) = N*A(I,J) + A(J,N-I+1) + 1
```

```
00530 :    NEXT J
00540 :  NEXT I
00550 :
00560 REM === DISPLAY=======================================
00570 :
00580 : PRINT
00590 : FOR I = 1 TO N
00600 :    LET S = 0
00610 :    FOR J = 1 TO N
00620 :       PRINT TAB(5*J) Q(I,J);
00630 :       LET S = S + Q(I,J)
00640 :    NEXT J
00650 :    PRINT TAB(5*(N+1)) "---";S
00660 :    PRINT
00670 : NEXT I
00680 :
00690 : FOR J = 1 TO N : PRINT TAB(5*J) "-----";: NEXT J
00700 :
00710 : PRINT
00720 : FOR J = 1 TO N
00730 :    LET S = 0
00740 :    FOR I = 1 TO N
00750 :       LET S = S + Q(I,J)
00760 :    NEXT I
00770 :    PRINT TAB(5*J) S;
00780 : NEXT J
00790 :
00800 : PRINT : PRINT
00810 :
00820 : END
```

****

It is much harder to construct even order magic squares than odd order ones. There are fancy mathematical formulas for doing so but let's try a different procedure here.

```
00100 : PRINT CHR$(147)
00110 : PRINT "         MAGIC SQUARE
00120 : PRINT "         ------------
00130 :
00140 REM CONSTRUCTS MAGIC SQUARES OF EVEN ORDER
00150 :
00160 REM === INPUT AND INITIALIZATION
00170 :
00180 : PRINT
00190 : INPUT "SIDE LENGTH (EVEN) "; N
00200 : IF N < 4 OR N > 20 OR 2*INT(N/2) <> N THEN 190
00210 :
00220 : LET H = N/2 : REM HALF SIDE LENGTH
00230 :
```

```
00240 : DIM Q(N,N), C(N,H)
00250 :
00260 : FOR I = 1 TO N : FOR J = 1 TO H : LET C(I,J) = 0
00265 : NEXT J : NEXT I
00270 :
00280 REM === CONSTRUCTION =====================================
00290 :
00300 : REM --- SETTING THE NUMBERS, MAGIC SUM OF THE ROWS
00310 :
00320 :    LET X = 0
00330 :
00340 :    FOR I = 1 TO N
00350 :      FOR J = 1 TO H
00360 :        LET X = X+1 : LET Q(I,J) = X
00370 :      NEXT J
00380 :      FOR J = H+1 TO N
00390 :        LET X = X+1 : LET Q(N-I+1,J) = X
00400 :      NEXT J
00410 :    NEXT I
00420 :
00430 : REM --- MAGIC SUM OF DIAGONALS
00440 :
00450 :    LET J = H
00460 :    FOR I = H TO 1 STEP -1
00470 :      LET J = J+1
00480 :      LET V = Q(I,J) : LET Q(I,J) = Q(N-I+1,J)
00485 :      LET Q(N-I+1,J) = V
00490 :    NEXT I
00500 :
00510 :    LET I = H-1
00520 :    FOR J = 1 TO H
00530 :      LET I = I+1
00540 :      IF I > H THEN LET I = 1
00550 :      LET V = Q(I,J) : LET Q(I,J) = Q(N-I+1,J)
00555 :      LET Q(N-I+1,J) = V
00560 :    NEXT J
00570 :
00580 : REM --- MAGIC SUM OF COLUMNS
00590 :
00600 :    LET T = H
00610 :    IF H = 2*INT(H/2) THEN 700 : REM N IS DIVISIBLE
00615 :                                  REM BY 4
00620 :
00630 :    LET I = H-1 : LET T = T-1
00640 :    FOR J = 1 TO H
00650 :      LET I = I+1
00660 :      IF I > H THEN LET I = 1
00670 :      LET C(I,J) = 1
00680 :    NEXT J
00690 :
00700 :    LET T = T/2
00710 :    FOR K = 1 TO T
```

264

```
00720 :     LET I = K
00730 :     FOR J = 1 TO H
00740 :       LET I = I+1
00750 :       IF I > H THEN LET I = 1
00760 :       LET C(I,J) = 1
00770 :       LET C(N-I+1,J) = 1
00780 :     NEXT J
00790 :   NEXT K
00800 :
00810 :   FOR I = 1 TO N
00820 :     FOR J = 1 TO H
00830 :       IF C(I,J) <> 1 THEN 850
00840 :       LET V = Q(I,J) : LET Q(I,J) = Q(I,N-J+1)
00845 :       LET Q(I,N-J+1) = V
00850 :     NEXT J
00860 :   NEXT I
00870 :
00880 REM === DISPLAY =======================================
00890 :
00900 : REM --- ROWS
00910 :
00920 :   PRINT
00930 :   FOR I = 1 TO N
00940 :     LET S = 0
00950 :     FOR J = 1 TO N
00960 :       PRINT TAB(5*J) Q(I,J);
00970 :       LET S = S + Q(I,J)
00980 :     NEXT J
00990 :     PRINT TAB(5*(N+1)) "---"; S
01000 :     PRINT
01010 :   NEXT I
01020 :
01030 :   PRINT CHR$(145);
01040 :   FOR J = 1 TO N : PRINT TAB(5*J) "-------";
01045 :   NEXT J
01050 :
01060 : REM --- COLUMN SUMS
01070 :
01080 :   PRINT
01090 :   FOR J = 1 TO N
01100 :     LET S = 0
01110 :     FOR I = 1 TO N : LET S = S + Q(I,J) : NEXT I
01120 :     PRINT TAB(5*J) S;
01130 :   NEXT J
01140 :
01150 : REM --- DIAGONAL SUMS
01160 :
01170 :   LET S1 = 0 : LET S2 = 0
01180 :   FOR I = 1 TO N
01190 :     LET S1 = S1 + Q(I,I)
01200 :     LET S2 = S2 + Q(I,N-I+1)
01210 :   NEXT I
```

```
01220 :
01230 :    PRINT "***"; S1; S2
01240 :
01250 :    END
```

Here is an explanation of the program. First the square Q
is filled with the numbers $1, 2, 3, \ldots N^2$ in such a way that
the totals of the rows are the same as the magic constants
(111 when n = 6). This happens in lines 320-410 with the
following result:

1	2	3	34	35	36
7	8	9	28	29	30
13	14	15	22	23	24
19	20	21	16	17	18
25	26	27	10	11	12
31	32	33	4	5	6

Now the diagonals have to be made magic, but in a way that
won't change the row totals (lines 450-560). In doing this,
the right parts of the main diagonal and the other diagonal
are exchanged and this is compensated for on the left side.
Here it is:

1	32	3	34	35	6
7	8	27	28	11	30
19	14	15	16	23	24
13	20	21	22	17	18
25	26	9	10	29	12
31	2	33	4	5	36

Handling the columns is the most difficult part (lines
600-860). First the C matrix is constructed, 1's occupying
Q matrix indices which are to be exchanged. (The C matrix
is built in lines 630-790, and the exchanging carried out in
lines 810-860. Only the left half is necessary for this
because the exchanges are done symmetrically with respect to
the middle. Perhaps some interested and motivated readers
can come up with the principle behind this process by dint
of their own thinking and experimenting.

BRAINTEASERS

1. Show that the famous magic square in Duerer's engraving "Melancholia" cannot be constructed by the Euler method. Here it is:

16	3	2	13
5	10	11	8
9	6	7	12
4	15	14	1

2. Mathematicians owe to Cornelius Agrippa (Agrippa von Nettesheim, 1486-1535) a method for the construction of magic squares of odd order. When $N = 5$ the result looks like this:

11	24	7	20	3
4	12	25	8	16
17	5	13	21	9
10	18	1	14	22
23	6	19	2	15

Expand this square to make a large checkerboard. Figure out the principle behind its construction and write the program.

3. Show that the following rules apply to magic squares M, M_1, M_2:

$$M = a + M_1$$
$$M = a * M_1$$
$$M = M_1 + M_2$$

where a is a positive integer. (It is assumed that addition and multiplication of a number and a matrix is component-wise).

4. The object is to insert the numbers $1, 2, 3, \ldots, 10$ into the ten circles on the figure pictured here so that the

total of the six numbers that make up the sides of the three smaller triangles a, b, and c is the same for each triangle. Write a program for this.

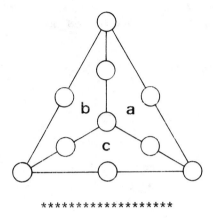

****

The following classic problem is as instructive as it is famous.

Example 8.2 TOO MANY QUEENS
Take a chess board measuring NxN and place N queens on it so that they do not threaten each other, which is to say that no queen may be in the same row, column, or diagonal.

Even the great mathematician Carl Friedrich Gauss (1777-1855) pondered this problem for the case N = 8. Naturally, he was interested not only in one, but all solutions. He sought an algorithm that would generate these solutions. The Gauss algorithm is as follows: "Move from left to right and place a queen in each column of the chess board -- always in the square closest to the top of the board. When you can't place another queen on a square, then move the queen in the immediately preceding column as far down as necessary until it is possible to continue."

In order to execute this algorithm with a program, we have to be able to show the arrangement of queens.

268

	1	2	3	4	5	6	7	8
1	O	·	·	·	·	·	·	·
2	·	·	·	·	·	·	O	·
3	·	·	·	·	O	·	·	·
4	·	·	·	·	·	·	·	O
5	·	O	·	·	·	·	·	·
6	·	·	·	O	·	·	·	·
7	·	·	·	·	·	O	·	·
8	·	·	O	·	·	·	·	·

We can express the solution pictured above in the following table:

Column i	1	2	3	4	5	6	7	8
Row d(i)	1	5	8	6	3	7	2	4

A shorthand version is the permutation 1 5 8 6 3 7 2 4. This is the form the computer uses to show the solutions on the screen. The Gauss method, with N = number of queens, goes like this:

TOO MANY QUEENS

```
column:= 1
row(column):= 0
WHILE column > 0   REPEAT
   REPEAT
      row(column):= row(column) + 1
   UNTIL not threatened OR row(column) > N
         END REPEAT
    IF row(column) > N THEN
       column:= column - 1   (*move back*)
    BUT IF column = N THEN
       display      (*solution found*)
       column:= column - 1
   OTHERWISE
   column:= column + 1 (*new queen*)
   row(column):= 0
   END IF
END REPEAT
```

Instead of writing "d(i)" we have used the term "row(column)." If we were using the programming language PASCAL, we could translate this algorithm directly into a program. BASIC, however, obscures this clear structure.

Before writing the program it is important to define what it means for a queen to be "threatened." The queens are placed in the same row when d(i) = d(k). We use i and k again for the column index. The queens are in the same diagonal if the line connecting them slopes at the rate of ±1. This rise is represented by the term

$$\frac{d(i) - d(k)}{i - k}$$

Because i > k we can also write a criterion for "threat" as follows: $|d(i) - d(k)| = i - k$. The test for a queen being threatened goes in a small subroutine.

```
00100 : PRINT CHR$(147)
00110 : PRINT "        TOO MANY QUEENS
00120 : PRINT "        ---------------
00130 :
00140 REM THE OBJECT IS TO PLACE N QUEENS ON A CHESS BOARD
00150 REM MEASURING NXN SO THEY DO NOT THREATEN EACH OTHER
00160 :
00170 REM === INPUT AND INITIALIZATION =====================
00180 :
00190 : PRINT : INPUT "HOW MANY QUEENS"; N
00200 : DIM D(N)
00210 : FOR I = 1 TO N : LET D(I) = 0 : NEXT I
00220 :
00230 REM === SEARCH =======================================
00240 :
00250 : LET I = 1
00260 : LET D(I) = 0
00270 :
00280 : LET D(I) = D(I) + 1          : REM QUEEN ADVANCES
00290 : GOSUB 430                     : REM TEST FOR THREAT
00300 : IF B$ = "THREATENED" AND D(I) <= N THEN 280
00305 :                                 REM REPEAT
00310 :
00320 : IF D(I) > N THEN 390         : REM RETREAT
00330 :
00340 : IF I < N THEN LET I = I+1 : LET D(I) = 0 : GOTO 280
00350 :
00360 : PRINT
00370 : FOR J = 1 TO N : PRINT D(J); : NEXT J
```

```
00375 :                              REM SOLUTION FOUND
00380 :
00390 : LET I = I-1 : IF I > 0 THEN 280 : REM CONTINUE
00395 :                              REM SEARCH
00400 :
00410 : END
00420 :
00430 REM +++ SUBROUTINE 'TEST FOR THREAT" +++++++++++++++++
00440 :
00450 : LET B$ = "NOT THREATENED"
00460 : IF I = 1 THEN RETURN : REM SPECIAL CASE
00470 : FOR K = 1 TO I-1
00480 :    LET A = ABS(D(I)-D(K))
00490 :    IF A=0 OR A = I-K THEN B$ = "THREATENED" : RETURN
00500 : NEXT K
00510 :
00520 : RETURN
00530 :
```

When N = 5 the program comes up with the following solutions:

```
1   3   5   2   4
1   4   2   5   3
2   4   1   3   5
2   5   3   1   4
3   1   4   2   5
3   5   2   4   1
4   1   3   5   2
4   2   5   3   1
5   2   4   1   3
5   3   1   4   2
```

There is a total of 92 possible solutions using 8 queens.
The first ones are as follows:

```
1   5   8   6   3   7   2   4
1   6   8   3   7   4   2   5
1   7   4   6   8   2   5   3
1   7   5   8   2   4   6   3
2   4   6   8   3   1   7   5
2   5   7   1   3   8   6   4
2   5   7   4   1   8   6   3
2   6   1   7   4   8   3   5
```

```
2  6  8  3  1  4  7  5
2  7  3  6  8  5  1  4
2  7  5  8  1  4  6  3
2  8  6  1  3  5  7  4
3  1  7  5  8  2  4  6
3  5  2  8  1  7  4  6
3  5  2  8  6  4  7  1
3  5  7  1  4  2  8  6
3  5  8  4  1  7  2  6
3  6  2  5  8  1  7  4
3  6  2  7  1  4  8  5
3  6  2  7  5  1  8  4
3  6  4  1  8  5  7  2
3  6  4  2  8  5  7  1
3  6  8  1  4  7  5  2
3  6  8  1  5  7  2  4
3  6  8  2  4  1  7  5
3  7  2  8  5  1  4  6
3  7  2  8  6  4  1  5
3  8  4  7  1  6  2  5
4  1  5  8  2  7  3  6
```

Wouldn't it be nice to able to watch the computer go through the search procedure on the screen? The following program makes it possible.

```
00100 : PRINT CHR$(147)
00110 : PRINT "      TOO MANY QUEENS
00120 : PRINT "      ---------------
00130 :
00140 REM SEARCH PROCESS SIMULTANEOUSLY DISPLAYED ON SCREEN
00150 :
00160 REM === INPUT AND INITIALIZATION ====================
00170 :
00180 : PRINT : INPUT "HOW MANY QUEENS"; N
00190 : IF N < 2 OR N > 10 THEN 180
00200 :
00210 : DIM D(N)
00220 : FOR I = 1 TO N : LET D(I) = 0 : NEXT I
00230 :
00240 : DEF FN D(X) = 32768-1 + 2*X + 80*D(X)
00250 :
00260 : LET D = 81 : REM CODE NUMBER OF THE QUEEN'S
```
272

```
00265 :                 REM CHARACTER
00270 :
00280 REM === GAME BOARD ==================================
00290 :
00300 : PRINT CHR$(147)
00310 : FOR W = 1 TO N : PRINT " -";: NEXT W : PRINT
00320 : FOR I = 1 TO N
00330 :   FOR J = 1 TO N : PRINT "| ";: NEXT J : PRINT "|"
00340 :   FOR W = 1 TO N : PRINT " -";: NEXT W : PRINT
00350 : NEXT I
00360 : PRINT
00370 :
00380 REM === SEARCH ======================================
00390 :
00400 : REM --- START POSITION
00410 :
00420 :   LET I = 1
00430 :   LET D(I) = 0
00440 :
00450 : REM --- QUEEN ADVANCES
00460 :
00470 :   LET D(I) = D(I)+1          : REM QUEEN MOVES
00480 :   POKE FND(I),D              : REM SET CHARACTERS
00490 :   IF D(I) > 1 THEN POKE FND(I)-80,32 : REM ERASE
00495 :                                       REM CHARACTERS
00500 :
00510 : REM --- TEST FOR THREAT
00520 :
00530 :   LET B$ = "NOT THREATENED"
00540 :   IF I = 1 THEN 620 : REM SPECIAL CASE
00550 :   FOR K = 1 TO I-1
00560 :     LET A = ABS(D(I)-D(K))
00570 :     IF A = 0 OR A = I-K THEN LET B$ = "THREATENED"
00580 :   NEXT K
00590 :
00600 :  IF B$ = "THREATENED" AND D(I) <= N THEN 450
00605 :                                  REM QUEEN ADVANCES
00610 :
00620 : REM --- QUEEN AT THE EDGE?
00630 :
00640 :   IF D(I) > N THEN 800 : REM RETREAT
00650 :
00660 : REM --- PLACE QUEEN IN NEXT COLUMN
00670 :
00680 :   IF I < N THEN LET I = I+1 : GOTO 430
00690 :
00700 : REM --- DISPLAY SOLUTION
00710 :
00720 :   PRINT CHR$(19) : FOR K = 1 TO 2*N : PRINT : NEXT K
00730 :   PRINT : FOR J = 1 TO N : PRINT D(J);: NEXT J
00740 :
00750 : REM --- CONTINUE SEARCH?
```

```
00760 :
00770 :    PRINT : PRINT "PRESS KEY TO CONTINUE SEARCH!
00780 :    GET T$ : IF T$ = "" THEN 780
00790 :
00800 : REM --- RETREAT
00810 :
00820 :    POKE FND(I),32     : REM ERASE QUEEN
00830 :    LET I = I-1        : REM TO THE COLUMN IN FRONT
00840 :    IF I > 0 THEN 450  : REM QUEEN ADVANCES AGAIN
00850 :
00860 : REM --- FINISHED
00870 :
00880 :    PRINT : PRINT "ALL THE SOLUTIONS HAVE BEEN FOUND.
00890 :
00900 :    END
```
****

BRAINTEASERS

5. The previous program for the graphic representation of solutions to the "Too Many Queens" has a little flaw in it. It briefly lets the queens go over the lower border of the chess board. Correct this detail.

6. By turning or reversing the position of the chess board, it is possible to derive all the solutions to the "Too Many Queens" from a smaller sample. Prove that the "8 Queen Problem" only has these 12 basic solutions:

15863724	16837425	24683175	25713864	25741863
26174835	26831475	27368514	27581463	35281746
	35841726	36258174		

Customize the program so that it produces only the basic solutions.

7. The humanist and mathematician Pierre de Fermat (1601-1655) formulated the "Two Square Theorem": "Every prime number having the form $p = 4*n + 1$ can be represented as the sum of two squares in only one way." Examples: 5 = 1+4; 13 = 4+9; 17 = 1+16; 29 = 4+25.... The theorem was first proven by Leonhard Euler (1707-1783) but not long ago C. Larson found a new proof for it that shows an affinity to

274

the "N Queen Problem". Larson used solutions such as the following:

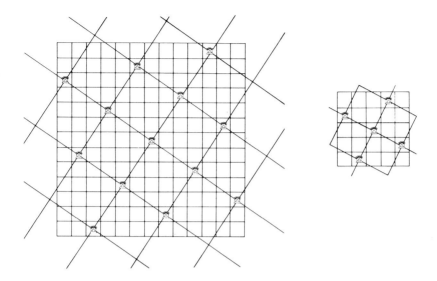

He showed that these solutions apply only when the side length is a prime number p of the form 4*n + 1. Each of the smaller squares (having queens in the corners) has the side length $\sqrt{p}$. This is the hypoteneuse of a right trangle having the sides a and b (of a generalized knight's move). Therefore, $p = a^2 + b^2$ and can be represented as the sum of two squares. Presumably old Papa Gauss, who scratched his head over these problems for a long time, made a break-through that led Larson to his conclusion.

Write a program to generate the first 100 prime numbers of the form p = 4*n + 1. Give the break-down of the square roots (a and b) and find the appropriate solutions to the queen problem.

8. SUPERQUEENS are chess pieces that can move both as queens and as knights. Write a program for positioning Superqueens on a board measuring N x N so they do not threaten each other.

9. N playing pieces are to be placed on a board measuring N x N so that precisely one piece is in each column, one in each row, and one in each diagonal. In doing this consider only those solutions as different that are not the result of turning or repositioning the board. When N = 4 there is only one solution: 1342; for N = 5 there are four: 13524, 14532, 21354, 25314.

****

The knight is sometimes called the comedian of the chess game. No other figure presents so many combinational problems. The oldest one is called the "Knight Errant." The game of chess uses 32 figures -- just the right number to fill half the squares on the chess board. This relationship inspired medieval chess players and theoreticians to pose the following problem: imagine half the board filled with figures and then try to capture each one with the knight in uninterrupted succession. This gives rise to the next problem of not only capturing half the board, but the whole one.

Example 8.3   KNIGHT ERRANT
The object is to construct a complete set of knight's moves on a board measuring N x N. The knight is not allowed to land on any one square more than once.

Examples:

1	12	23	18	7
22	17	8	13	24
11	2	25	6	19
16	21	4	9	14
3	10	15	20	5

28	23	6	15	34	21
7	16	27	22	5	14
24	29	8	35	20	33
9	36	17	26	13	4
30	25	2	11	32	19
1	10	31	18	3	12

When N = 6 the knight's move is closed which means that from
the end square (move 36) the knight jumps back to the beginning square (move 1).

Now let's develop a program that will create these
moves. The chess board is represented by a two-dimensional
table b(,); b(i,j) gives us the contents of the board in
row i and column j.

	1	2	3	4	5	6	7	8
1	.	.	.	.	.	.	.	.
2	.	.	6	.	7	.	.	.
3	.	5	.	.	.	8	.	.
4	.	.	.	O	.	.	.	.
5	.	4	.	.	.	1	.	.
6	.	.	3	.	2	.	.	.
7	.	.	.	.	.	.	.	.
8	.	.	.	.	.	.	.	.

We store the vectors of the 8 possible moves in two tables
si( ), and sj( ). Note that si(1) = 1, sj(1) = 2, meaning
that the knight moves 1 square down and 2 squares to the
right. Furthermore, move number 2 above is represented by:
si(2) = 2, sj(2) = 1. Move number 3 is si(3) = 2, sj(3) =
-1, etc.

We'll let the knight start out in square (1,1). He
moves to square (2,3) and then to (3,5), etc. He always
tests every square he wants to occupy to see if he is allowed there. He has to know if he has visited it before,
whether it lies outside the board's dimensions, etc. If the
square is not open to him, then the knight tries the next
move among the set of 8 possible ones he has. If none of
the 8 is possible, he is trapped and cannot move. We'll
discuss his options in a moment, but first, the program:

```
00100 : PRINT CHR$(147)
00110 : PRINT "          KNIGHT ERRANT
00120 : PRINT "          ------------
00130 :
00140 REM KNIGHT MOVES ON CHESS BOARD MEASURING N X N
00150 REM UNTIL IT ENDS IN A CUL-DE-SAC
00160 :
00170 REM === INPUT AND INITIALIZATION ====================
00180 :
00190 : REM --- CHESS BOARD AND STARTING POSITION
00200 :
00210 :    PRINT : INPUT "SIDE LENGTH OF CHESS BOARD"; N
00220 :
00230 :    DIM B$(N,N)
00240 :    FOR I = 1 TO N
00250 :      FOR J = 1 TO N : LET B$(I,J) = "." : NEXT J
00260 :    NEXT I
00270 :
00280 :    PRINT : INPUT "BEGINNING COORDINATES "; I0, J0
00290 :
00300 : REM --- COORDINATES OF KNIGHT'S MOVE
00310 :
00320 :    FOR K = 1 TO 8 : READ SI(K), SJ(K) : NEXT K
00330 DATA 1,2,2,1,2,-1,1,-2,-1,-2,-2,-1,-2,1,-1,2
00340 :
00350 REM === SERIES OF MOVES ============================
00360 :
00370 : REM --- BEGINNING COORDINATES
00380 :
00390 :    LET I = I0 : LET J = J0
00400 :    LET Z = 1  : REM NUMBER OF INDIVIDUAL MOVE
00410 :
00420 : REM --- MOVE NUMBER Z
00430 :
00440 :    LET B$(I,J) = STR$(Z)
00450 :
00460 : REM --- NEXT POSSIBLE MOVE
00470 :
00480 :    LET Z = Z + 1
00490 :
00500 :    FOR K = 1 TO 8
00510 :      LET I1 = I + SI(K) : LET J1 = J + SJ(K)
00520 :      IF I1 < 1 OR I1 > N THEN 540
00525 :      IF J1 < 1 OR J1 > N THEN 540
00530 :      IF B$(I1,J1) = "." THEN I = I1 : J=J1 : GOTO 420
00540 :    NEXT K
00550 :
00560 REM === DISPLAY ===================================
00570 :
00580 : PRINT
```

278

```
00590 : FOR R = 1 TO N
00600 :   FOR S = 1 TO N
00610 :     PRINT TAB(4*S);
00620 :     IF B$(R,S) = "." THEN PRINT B$(R,S); : GOTO 640
00630 :     PRINT CHR$(157) B$(R,S);
00640 :   NEXT S
00650 :   PRINT
00660 : NEXT R
00670 :
00680 : END
```

This program generates the following pattern of moves:

1	•	•	•	•	•	•	•
•	•	2	•	•	•	•	•
•	32	•	•	3	•	•	•
34	•	•	27	•	•	4	•
31	10	33	•	•	26	23	•
18	35	28	11	24	21	14	5
9	30	19	16	7	12	25	22
36	17	8	29	20	15	6	13

After move 36 the knight is cornered down in the lower left of the board. Because he was always moving clock-wise to search for possible moves, he almost never left the lower half of the board. Can we avoid this counter-productive tendency?

Now what happens if we let the knight determine his moves at random? This eliminates any preference for one direction and will give his travels a completely different pattern. Here are two possibilities:

1	•	3	•	13	•	25	•
4	•	•	•	•	21	14	23
•	2	•	12	7	24	•	26
•	5	•	•	20	15	22	•
•	•	19	6	11	8	27	16
32	•	•	35	18	41	10	39
•	36	33	30	9	38	17	28
•	31	•	37	34	29	40	•

1	4	19	24	17	•	•	•
48	23	2	5	20	25	16	•
3	8	21	18	37	•	•	26
22	47	6	•	•	15	38	35
7	•	9	46	39	36	27	14
10	45	42	•	12	31	34	29
43	•	11	40	•	28	13	32
•	41	44	•	•	33	30	•

In these examples the knight made more progress than when his moves were predetermined. Nonetheless, he ultimately was blocked. His mistake is apparent: he incautiously ignored several squares on the edge and in the corners of the board. Later it was impossible to visit these again.

This is an observation that can help us develop the moves in Knight Errant. This rule is named after the mathematician J.C. Warnsdorff (1823) and goes like this: "In choosing the knight's move, examine the squares open to him from each possible new position. Choose from these the one that permits the fewest moves to unoccupied squares." Here we foresee the danger of not being able to reach the square in question again, thereby unintentionally omitting it altogether. The practical usefulness of this rule is so great that it applies not only to the arbitrary start of the knight's moves but also toward the end of the game if the player has been ignoring the rule all along. To program this rule we have to assign a value to every square of the board. This will indicate how many squares the knight can reach in each move. The square-values of a board measuring 8 x 8 are:

2	3	4	4	4	4	3	2
3	4	6	6	6	6	4	3
4	6	8	8	8	8	6	4
4	6	8	8	8	8	6	4
4	6	8	8	8	8	6	4
4	6	8	8	8	8	6	4
3	4	6	6	6	6	4	3
2	3	4	4	4	4	3	2

Note the high degree of symmetry to this pattern! Knowing only the values of the 10 squares in the triangle will be enough. The knight will now choose from among those squares available to it, that one with the lowest value. Once it has landed on a square, then the value of all the squares reachable in one jump is lowered by a factor of 1. Hope you're feeling ambitious. The program is left to you.

****

BRAINTEASERS

10. Prove the following problems:

   a) The smallest chessboard on which a complete game of Knight Errant can be played has side length 5.

   b) The smallest chessboard on which a complete and closed game of Knight Errant is possible has side length 6.

11. Write a program that uses the Warnsdorff Rule to generate knights' movements as in the game Knight Errant. Let the knight make about 20 random moves. After that point have the program apply the Warnsdorff Rule.

12. A further possibility for altering the knight's movement is suggested by the algorithm used to solve Too Many Queens: backtrack via the same moves that led the knight to the dead end. Try to program this option.

13. The game of Knight Errant doesn't necessarily have to be played on a square chess board. Try it with a rectangle or another geometric shape.

14. Euler was able to construct a game of Knight Errant with a special kind of board: it was also a (half) magic square. However, the diagonals don't add up. Try something similar.

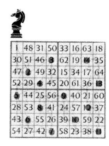

281

Before the turn of the century a popular game of strategy and concentration captured the interest of players worldwide. Compared to the sensation caused by this little game, the Rubik's Cube craze of our time is nothing. The game is called the "Fifteen Blocks" and is still popular today. Sam Lloyd, whose creative imagination has inspired portions of this book, is credited with the game's invention. Soon after the game's invention in 1878 the puzzle fever had spread throughout the world. So many people were playing it so eagerly that in some countries offices banned it during working hours.

Example 8.4 FIFTEEN BLOCKS
Fifteen moveable blocks, each numbered from 1 through 15, are enclosed in a frame. One empty space makes it possible to move the pieces around. The object is to rearrange the pieces so that they appear in natural numerical order.

There are 20,922,789,888,000 different starting positions. Half of these have a solution, half do not. When Sam Lloyd invented the game, he offered 100 dollars to anyone who could get to the winning position when the pieces were arranged as follows:

1	2	3	4
5	6	7	8
9	10	11	12
13	15	14	

The inventor's money was safe. It is impossible to get to the winning position from this arrangement. With the help of permutation theory we can derive a simple criterion for the situations in which a starting position has a solution.

We want to develop a program that will generate the final winning position from any starting position, if it exists. For the sake of simplicity we are going to limit our game to only 8 pieces sliding in a frame measuring 3 x 3. Unlike the programs for constructing magic squares or for the knight's move, here we are going to develop a pure search mechanism. This means having to examine all possibilities of combinations that result from further combinations. For example:

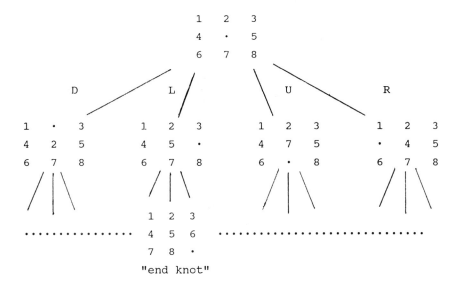

"end knot"

Here is the key to the diagram:

     D.......down (piece is moved down to empty space)
     L.......left (piece moved left to empty space)
     U.......up (piece moved up to empty space)
     R.......right (piece moved right to empty space)

This kind of diagram is called a tree. The individual positions may be called the knots. Our task is to search through the tree for the winning knot. The search process for examining this tree has the following components:

(1) Two lists: first, the list of "knots already examined" (containing all those whose successive knots have been identified); second, the list of "knots not yet examined" (this contains the knots whose successors haven't yet been identified.)

(2) A way to derive all successors of a particular knot (a

successor is a knot derived from the given knot by the operations D, L, U, or R).

(3)   A test to tell whether a knot that has been generated is really a winning knot.

(4)   A way of selecting the knot to be examined next.

TREE SEARCH

```
Add the beginning knot to the list of unexamined knots
REPEAT
   A:= uppermost element in the list of unexamined knots
   Add A to the list of examined knots
   Determine the successor of A
   IF a successor exists THEN
      FOR every successor to A REPEAT
         test for winning knot
         IF winning knot has been found THEN
            z:= Number of the winning knot
            Z$:= "Goal reached"
            END REPEAT
         OTHERWISE
            calculate the value of the knot
            add the knot to the list of unexamined knots
         END IF
      END REPEAT
   END IF
UNTIL list of unexamined knots is empty END REPEAT
IF Z$:= "Goal reached" THEN
   Determine the path to the solution by backtracking
   Display the solution
OTHERWISE
   Display: "There is no solution."
END IF
```

How do we represent the knots of the search tree in the program? In order to store these and be able to work with them easily, it is best to show them as sequences of characters, for example:

284

```
                          1    .    2
                          4    6    3
                          7    5    8
```

are going to combine to become 1.2463758. Two more bits of
information may be stored in this sort of character sequence. The first is a pointer to the preceding knot (necessary for finding it when back-tracking); and second, the
direction of the move that led from the preceding knot to
the knot the program is now examining. In addition, every
knot K$(NR) is given an order value O(NR) that determines
whether our knot is the next to be examined (to "examine" a
knot means in this sense to identify all its successors).
Here is an example: the character sequence

        K$(10) = 007   L   1.2463758   with   O(10) = 5

expresses that the configuration   1.2
                                    463
                                    758   had the predecessor
no.7 and was derived from this by move L.   Its order value
is 5.

     The previously examined knots are given a value >
90,000. This trick prevents them from being examined again.
One of a knot's successors is eliminated immediately, namely
the one that corresponds to the predecessor of the examined
knot.   This is tested in lines 2260-2320 of the following
program.

     The program has been written so that it should document
itself.   By building in this capability it is a bit slower
than usual.   In order to speed things up, a user may remove
the comments.   Furthermore, in order to leave a few finishing touches for the user to add, the program has been left
incomplete.   You lack the subroutine that proceeds from the
target knot and reconstructs the order of moves that led
from the starting knot to the target knot (line 3660 ff.)
Line 1230 incorporates the search strategy.   Users should
experiment a little if they want to explore this aspect of

the program. You'll find more suggestions in the BRAIN-
TEASERS that follow.

```
00100 : PRINT CHR$(147)
00110 : PRINT "           FIFTEEN BLOCKS
00120 : PRINT "           --------------
00130 :
00140 REM SEARCH PROGRAM DESIGNED TO RECONSTRUCT BASIC
00150 REM CONFIGURATION (IF POSSIBLE) FROM A
00155 REM GIVEN STARTING POSITION
00160 :
00170 REM === INITIALIZATION AND DISPLAY ===================
00180 :
00190 : REM --- TABLES
00200 :
00210 :    DIM K$(100) : REM LIST OF KNOTS
00220 :    DIM O(100)  : REM TABLE OF ORDER VALUES
00230 :    DIM NF$(4)  : REM TABLE OF SUCCESSORS TO A KNOT
00240 :    DIM E$(3,3) : REM GAME FIELD
00250 :    DIM F$(3,3) : REM HELPING GAME FIELD
00260 :
00270 : REM --- INCREMENTS FOR THE MOVEMENT OF THE POINT
00280 :
00290 :    LET X(1) = -1 : LET Y(1) = 0
00300 :    LET X(2) = 0 : LET Y(2) = 1
00310 :    LET X(3) = 1 : LET Y(3) = 0
00320 :    LET X(4) = 0 : LET Y(4) = -1
00330 :
00340 : REM --- CHARACTER SERIES FOR DIRECTION OF MOVE
00350 :
00360 :    LET R$ = "DLURB" : REM DOWN, LEFT, UP, RIGHT,
00365 :                        REM BEGIN
00370 :    LET I$ = "URDL"  : REM INVERSE DIRECTION
00380 :    LET ZR$ = "B" : REM DIRECTION OF MOVE (BEGINNING
00385 :                    REM VALUE 'BEGIN')
00390 :
00400 : REM --- ENTER THE STARTING POSITION
00410 :
00420 :    PRINT
00430 :    PRINT "ENTER THE STARTING POSITION IN THE FORM ";
00440 :    PRINT "'1234.5678'";
00450 :    INPUT A$
00460 :    IF LEN(A$) = 9 THEN 480
00465 :    PRINT "INPUT ERROR!" : GOTO 430
00470 :
00480 : REM --- INITIALIZATION OF THE VARIABLES
00490 :
00500 :    LET NR = 1    : REM NUMBER OF KNOTS
00510 :    LET NN = 1    : REM NUMBER OF KNOT DEVELOPED
00520 :    LET O(1) = 0  : REM ORDER VALUE OF FIRST KNOT
```

```
00530 :   LET BK = 0   : REM NUMBER OF EXAMINED KNOTS
00540 :
00550 : REM --- INITIALIZATION OF LIST OF UNEXAMINED KNOTS
00560 :
00570 :   GOSUB 2930 : REM KNOT NUMBER AS CHARACTER SERIES
00580 :
00590 :   LET K$(1) = NN$ + "B" + A$ : REM DESIGNATION OF
00595 :                                 REM STARTING KNOT
00600 :
00610 :   LET K$ = K$(1) : GOSUB 3200 : REM GOAL ALREADY
00615 :                                 REM REACHED?
00620 :   IF ZF$ <> "GOAL REACHED" THEN 640
00625 :   PRINT "WHAT DOES THAT MEAN?" : END
00630 :
00640 : REM --- INITIALIZATION OF END FLAG
00650 :
00660 :   LET EF$ = "NOT FINISHED"
00670 :
00680 REM === SEARCH =====================================
00690 :
00700 : REM DETERMINE UPPER ELEMENT OF THE UNEXAMINED LIST
00710 :
00720 :   GOSUB 1630 : REM DETERMINING THE KNOT WITH LOWEST
00725 :               REM VALUE
00730 :
00740 :   IF O(NN) < 90000 THEN 760
00745 :   LET EF$ = "FINISHED" : GOTO 1400
00750 :
00760 : REM --- PUT KNOTS IN LIST OF EXAMINED KNOTS
00770 :
00780 :   PRINT
00790 :   PRINT "KNOT NUMBER"; NN; "NAMED "; K$(NN);
00800 :   PRINT "WITH ORDER VALUE "; O(NN);" IS GENERATED
00810 :
00820 :   PRINT "SUCCEEDING KNOTS: "
00830 :
00840 :   LET BK = BK + 1     : REM NUMBER OF EXAMINED KNOTS
00850 :   LET O(NN) = O(NN) + 90000 : REM MARK AS EXAMINED
00860 :
00870 : REM --- DETERMINE THE SUCCESSORS
00880 :
00890 :   LET E$  = MID$(K$(NN),5,9) : REM RETRIEVE NUMBER
00895 :                               REM CONFIGURATION
00900 :   LET ZR$ = MID$(K$(NN),4,1) : REM DIRECTION OF MOVE
00910 :
00920 :   GOSUB 2000     : REM 'UNPACK' CHARACTER SERIES
00930 :   GOSUB 2180                 : REM DEVELOP KNOTS
00940 :
00950 : REM --- DOES A SUCCESSOR EXIST?
00960 :
00970 :   IF NF > 0 THEN 1020 : REM SUCCESSOR EXISTS
00980 :
```

287

```
00990 :    GOTO 1400          : REM END OF SEARCH?
01000 :
01010 :
01020 : REM --- LOOP FOR ALL LEGAL SUCCESSOR KNOTS
01030 :
01040 :    FOR R = 1 TO NF : REM START LOOP ----------------
01050 :
01060 :       LET K$ = NF$(R) : GOSUB 3200 : REM COMPARE WITH
01065 :                                      REM GOAL
01070 :
01080 :       REM --- COMPARE WITH GOAL KNOT
01090 :
01100 :          IF ZF$ <> "GOAL REACHED" THEN 1160
01110 :
01120 :       REM --- GOAL KNOT REACHED
01130 :
01140 :          LET EF$ = "FINISHED"  : GOTO 1400
01150 :
01160 :       REM --- GOAL KNOT NOT REACHED
01170 :
01180 :          LET E$ = MID$(NF$(R),5,9)
01190 :          GOSUB 2000 : REM UNPACK CHARACTER SERIES
01200 :
01210 :       REM --- ADD KNOTS TO LIST
01220 :
01230 :          LET RG = O(NN) + 1 - 90000 : REM SEARCH
01235 :                                       REM STRATEGY
01240 :
01250 :          LET NR = NR + 1 : REM RAISE CURRENT KNOT
01255 :                            REM NUMBER
01260 :          LET K$(NR) = NF$(R) : REM ADD KNOTS TO LIST
01270 :          LET O(NR) = RG     : REM NEW ORDER NUMBER
01280 :
01290 :    NEXT R : REM END LOOP ----------------------------
01300 :
01400 :    IF EF$ <> "FINISHED" THEN 700
01410 :
01420 :    GOTO 3410 : REM FOR DISPLAY
01430 :
01440 REM +++ SUBROUTINE 'IS THERE A NEW KNOT?' ++++++++++++
01450 :
01460 : REM INPUT: BOARD F$( , ), LIST K$( )
01470 : REM TEST TO SEE IF F$ HAS ALREADY BEEN EXAMINED
01480 : REM OUTPUT: DF$ = "DOUBLED" OR DF$ = "NOT DOUBLED"
01490 :
01500 : LET DF$ = "NOT DOUBLED"
01510 :
01520 : FOR I = 1 TO NR : REM RUN THROUGH ALL EXAMINED KNOTS
01530 :    IF O(I) < 90000 THEN 1550
01540 :    IF F$ <> MID$(K$(I),5,9) THEN 1550
01545 :    LET DF$ = "DOUBLED": RETURN
01550 : NEXT I
```

```
01560 :
01570 : IF DF$ = "DOUBLED" THEN PRINT "KNOTS DOUBLED"
01580 :
01590 : RETURN
01600 :
01610 REM END OF SUBROUTINE 'IS THERE A NEW KNOT?' +++++++++
01620 :
01630 REM SUBROUTINE DETERMINING KNOT OF LOWEST ORDER VALUE
01640 :
01650 : REM INPUT: LIST K$() OF KNOTS AND THEIR NUMBER NR
01660 : REM KNOT WITH LOWEST ORDER VALUE
01670 : REM OUTPUT: NN AND T, WHERE T = O(NN) IS THE LOWEST
01675 : REM VALUE
01680 :
01690 : LET T = 99999
01700 : LET N = 1
01710 :
01720 : FOR I = 1 TO NR
01730 :    IF T <= O(I) THEN 1760
01740 :    LET T = O(I)
01750 :    LET N = I
01760 : NEXT I
01770 :
01780 : LET NN = N : REM KNOT TO BE EXAMINED
01790 :
01800 : RETURN
01810 :
01820 REM END OF DETERMINING THE KNOT WITH LOWEST VALUE
01830 :
01840 REM SUBROUTINE PACK NUMBER FIELD INTO CHARACTER SERIES
01850 :
01860 : REM INPUT: BOARD F$( , )
01870 : REM TRANSFORM INTO CHARACTER SERIES
01880 : REM OUTPUT: F$ = CHARACTER SERIES OF 9 CHARACTERS
01890 :
01900 : LET F$ = ""
01910 :
01920 : FOR P = 1 TO 3
01930 :    FOR Q = 1 TO 3 : LET F$ = F$ + F$(P,Q) : NEXT Q
01940 : NEXT P
01950 :
01960 : RETURN
01970 :
01980 REM END SUBROUTINE NUMBER FIELD INTO CHARACTER SERIES
01990 :
02000 REM SUBROUTINE CHARACTER SERIES BECOMES NUMBER FIELD
02010 :
02020 : REM INPUT : CHARACTER SERIES E$
02030 : REM E$ GETS CHANGED TO NUMBER FIELD E$( , ),F$( , )
02040 : REM OUTPUT: NUMBER FIELDS E$( , ) AND F$( , )
02050 :
02060 : FOR I = 1 TO 3
```

```
02070 :    FOR J = 1 TO 3
02080 :       LET U = 3*(I-1) + J
02090 :       LET E$(I,J) = MID$(E$,U,1)
02100 :       LET F$(I,J) = E$(I,J)
02110 :    NEXT J
02120 : NEXT I
02130 :
02140 : RETURN
02150 :
02160 REM END OF SUBROUTINE  ++++++++++++++++++++++++++++++++
02170 :
02180 REM +++ SUBROUTINE 'DEVELOP KNOTS' ++++++++++++++++++
02190 :
02200 : REM ENTER: E$( , ) = F$( , ), KNOT NUMBER NN,
02205 : REM DIRECTION ZR$. AS MANY AS 3 PERMISSIBLE
02210 : REM SUCCESSOR KNOTS GENERATED. KNOTS ALREADY
02220 : REM DEVELOPED ARE ELIMINATED AND THE COMPLETE KNOT,
02230 : REM CONSISTING OF MOVE DIRECTION, POINTER AND BOARD,
02235 : REM IS CREATED
02240 : REM OUTPUT: LIST NF$( ) OF SUCCESSORS AND AMOUNT
02250 :
02260 : REM DETERMINING THE ILLEGAL DIRECTION OF DEVELOPMENT
02270 :
02280 :    FOR I = 1 TO 5
02290 :       IF ZR$ = MID$(R$,I,1) THEN 2320 : REM MOVE
02295 :                                         REM DIRECTION
02300 :    NEXT I
02310 :
02320 :    LET VR$ = MID$(I$,I,1) : REM ILLEGAL DIRECTION OF
02325 :                             REM DEVELOPMENT
02330 :
02340 : REM DETERMINING THE COORDINATES X1,Y1 OF THE EMPTY
02345 : REM FIELD
02350 :
02360 :    FOR Y1 = 1 TO 3
02370 :       FOR X1 = 1 TO 3
02380 :       IF E$(X1,Y1) = "." THEN 2420 : REM LOCATION HAS
02385 :                                      REM BEEN FOUND
02390 :      NEXT X1
02400 :    NEXT Y1
02410 :
02420 : REM --- GENERATING 4 POSSIBLE SUCCESSORS
02430 :
02440 :    LET AN = 0 : REM NUMBER OF SUCCESSORS AFTER TEST
02450 :    LET NF = 0 : REM NUMBER BEFORE TEST FOR DOUBLES
02460 :
02470 :    LET S1 = 1      : REM LOOP OF ALL SUCCESSORS
02480 :
02490 :       IF MID$(R$,S1,1) = VR$ THEN 2870 : REM SKIP
02495 :                                          REM PREDECESSOR
02500 :
02510 :       REM --- NEW LOCATION OF EMPTY FIELD
```
290

```
02520 :
02530 :        LET X2 = X1 + X(S1)
02540 :        LET Y2 = Y1 + Y(S1)
02550 :
02560 :        IF X2 < 1 OR X2 > 3 THEN 2870 : REM NOT LEGAL
02570 :        IF Y2 < 1 OR Y2 > 3 THEN 2870 : REM NOT LEGAL
02580 :
02600 :     REM --- INTERCHANGE THE FIELDS X1,Y1 AND X2,Y2
02610 :
02620 :        FOR I = 1 TO 3
02630 :          FOR J = 1 TO 3 : LET F$(I,J) = E$(I,J)
02635 :          NEXT J
02640 :        NEXT I
02650 :
02660 :        LET F$(X1,Y1) = F$(X2,Y2)
02670 :        LET F$(X2,Y2) = "."
02680 :
02690 :     REM --- KNOT NUMBER AS CHARACTER SERIES
02700 :
02710 :        GOSUB 2930 : REM KNOT NUMBER BECOMES CHARACTER
02715 :                        REM SERIES
02720 :
02730 :     REM --- TEST, WHETHER KNOT IS ALREADY PRESENT
02735 :     REM      IN THE TABLE
02740 :
02750 :        GOSUB 1440        : REM IS THERE A NEW KNOT?
02760 :        IF DF$ = "DOUBLED" THEN 2870 : REM TO THE NEXT
02765 :                                       REM KNOT
02770 :
02780 :     REM --- FORM COMPLETE CHARACTER SERIES
02790 :
02800 :        LET AN = AN + 1 : REM NUMBER OF SUCCESSORS
02805 :                             REM GENERATED
02810 :        GOSUB 1840    : REM REPACK AS CHARACTER SERIES
02820 :        LET NF$(AN) = NN$ + MID$(R$,S1,1) + F$
02830 :        LET NF = NF + 1
02840 :
02850 :        PRINT CHR$(157) AN; "   "; NF$(AN)
02860 :
02870 :   IF S1 < 4 THEN LET S1 = S1 + 1 : GOTO 2490
02880 :
02890 : RETURN
02900 :
02910 REM END OF SUBROUTINE 'DEVELOPS KNOTS' ++++++++++++++
02920 :
02930 REM SUBROUTINE 'KNOT NUMBER AS CHARACTER SERIES' +++++
02940 :
02950 : REM INPUT: NUMBER NN
02960 : REM MAKES NN INTO CHARACTER SERIES, INSERTS LEADING
02965 : REM ZEROES
02970 : REM OUTPUT: CHARACTER SERIES NN$ AS KNOT NUMBER
02980 :
```

```
02990 : LET NN$ = STR$(NN)
03000 : LET NN$ = RIGHT$(NN$,LEN(NN$)-1) : REM REMOVE
03005 :                                  REM LEADING SPACE
03010 : LET L = 3 - LEN(NN$)
03020 :
03030 : IF L < 0 THEN PRINT "RUN OVER ERROR IN 3030!" : END
03040 : IF L = 0 THEN RETURN
03050 :
03060 : FOR I = 1 TO L  : LET NN$ = "0" + NN$ : NEXT I
03070 :
03080 : RETURN
03090 :
03100 REM END OF 'KNOT NUMBER AS CHARACTER SERIES' +++++++++
03110 :
03200 REM SUBROUTINE 'TEST TO SEE IF TARGET KNOT REACHED'
03210 :
03220 : REM INPUT: KNOT K$
03230 : REM TAKES NUMBER CONFIGURATION OUT OF CHARACTER
03240 : REM SERIES AND COMPARES WITH TARGET BOARD
03250 : REM OUTPUT: ZF$ = "GOAL REACHED" OR "GOAL NOT
03255 : REM REACHED"
03260 :
03270 : LET ZF$ = "GOAL NOT REACHED"
03280 :
03290 : LET ZK$ = "12345678."   : REM TARGET KNOT
03300 : LET PK$ = RIGHT$(K$,9)  : REM KNOTS TO BE TESTED
03310 :
03320 : IF PK$ = ZK$ THEN LET ZF$ = "GOAL REACHED"
03330 :
03340 : RETURN
03350 :
03360 REM END OF TEST FOR TARGET KNOT ++++++++++++++++++++++
03370 :
03380 REM END OF SEARCH ====================================
03390 :
03400 :
03410 REM === DISPLAY =======================================
03420 :
03430 :    IF ZF$ <> "GOAL REACHED" THEN 3520
03440 :
03450 : REM --- DISPLAY PATH TO WINNING SOLUTION
03460 :
03470 :    PRINT : PRINT : PRINT "SOLUTION FOUND!
03480 :    GOSUB 3660 : REM RETRACE
03490 :    PRINT "THE WINNING PATH IS: ****** ";
03500 :    GOTO 3560
03510 :
03520 : REM --- NOTIFICATION OF GAME LOST
03530 :
03540 :    PRINT : PRINT "NO SOLUTION FOUND.
03550 :
03560 : REM --- DISPLAY THE KNOT AMOUNTS
```

```
03570 :
03580 :    PRINT : PRINT
03590 :    PRINT "NUMBER OF UNEXAMINED KNOTS: "; NR-BK
03600 :    PRINT "NUMBER OF EXAMINED KNOTS: "; BK
03610 :
03620 :    END
03630 :
03640 REM END OF DISPLAY ===================================
03650 :
03660 REM SUBROUTINE 'RETRACE THE PATH OF WINNING SOLUTION'
03670 :
03680 : REM INPUT: LIST OF POSSIBLE KNOTS K$
03690 : REM PATH IS RETRACED USING THE POINTER AND THE
03700 : REM DIRECTION LETTERS
03710 : REM OUTPUT: ORDER OF WINNING MOVES
03720 :
03730 : REM *** THE USER MUST ENTER THIS PORTION OF THE ***
03735 : REM                  *** PROGRAM ***
03740 :
03750 : RETURN
03760 :
03770 REM END OF SUBROUTINE 'RETRACE' ++++++++++++++++++++++
03780 :
```

****

BRAINTEASERS

16. Add the missing process to the previous program that will give the path to the solution via back-tracking.

17. Expand the program for a board measuring 4 x 4.

18. Prove that the following configuration of numbers has no solution:

1	2	3	4
5	6	7	8
9	10	11	12
13	15	14	.

19. Line 1230 of the Fifteen Blocks program contains the search strategy. Use examples to confirm that the tree is searched in the following way:

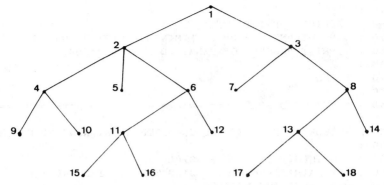

Change line 1230 to employ the following search strategy:

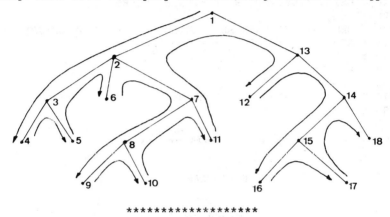

****

Here is an old puzzle that has been a favorite in magazines and childrens' books for at least three hundred years. A picture shows a number of containers (usually three), each holding a certain volume of water. The reader is then asked to figure out how many times one would have to pour the containers into each other to end up with specific amounts of water in various containers. It usually ends up that the largest one is filled. This is how the puzzle is stated in a work mentioned earlier, the Problemes des Bachet de Meziriac:

Example 8.5 THE POURING PUZZLE

Two friends have decided to divide a quantity of wine that they have stored in an eight-gallon cask. How can they pour the wine into two equal four gallon portions when they have

294

two smaller empty jugs holding three gallons and five gallons respectively. The jugs are not marked to show levels of liquid.

Bachet suggests two solutions:

$$800 \to 350 \to 323 \to 620 \to 602 \to 152 \to 143 \to 440$$
and
$$800 \to 503 \to 530 \to 233 \to 251 \to 701 \to 710 \to 413 \to 440$$

These numbers are interpreted as follows: 251 tells us that the first container (capacity 8) has 2 gallons in it; the second (capacity 5) holds 5; the third (capacity 3) holds 1.

With every pour one of the containers is either emptied into another or filled by another. Therefore, one of the containers has to be either completely full or completely empty. It is possible to create a whole series (or "cycle") of pours that ends with the same amounts in the various containers that you had at the beginning, and which under certain circumstances may contain all volumes from zero to the highest number. This series would contain the solution to the problem that required you to divide the number of gallons into two equal amounts.

The rule according to which the first solution is generated is as follows: "Fill from the largest container into the middle-sized one, from the middle-sized one to the smallest, and from the smallest into the largest. Repeat this procedure unless the next pour would result in a combination already seen; in this case, simply skip to the next part of the procedure."

```
00100 : PRINT CHR$(147)
00110 : PRINT "        POURING PUZZLE
00120 : PRINT "        --------------
00130 :
00140 REM A COMPUTER SOLUTION TO THE CLASSIC 3-CONTAINER
00145 REM PROBLEM
00150 :
00160 REM -- VARIABLES:
00170 :
00180 REM   G, M, S ...... CONTENTS OF THE LARGE, MEDIUM
00190 REM                  AND SMALL SIZED CONTAINERS
```

```
00200 REM    CG, CM, CS ... VOLUMES OF THE 3 VESSELS
00210 REM    BG, BM, BS ... BEGINNING VOLUMES OF 3 VESSELS
00220 REM    EG, EM, ES ... FINAL VOLUMES
00230 REM    N ........... NUMBER OF POURINGS
00240 :
00250 REM === ENTER AND INITIALIZATION =====================
00260 :
00270 : PRINT
00280 : PRINT "ENTER THE CAPACITIES OF THE 3 CONTAINERS!
00290 : INPUT "CAPACITY OF GREATEST CONTAINER   "; CG
00300 : IF CG <= 0 THEN 290
00310 : INPUT "CAPACITY OF MIDDLE CONTAINER      "; CM
00320 : IF CM <= 0 OR CM > CG THEN 310
00330 : INPUT "CAPACITY OF SMALL CONTAINER       "; CS
00340 : IF CS <= 0 OR CS > CM THEN 330
00350 :
00360 : PRINT
00370 : PRINT "ENTER THE BEGINNING VOLUMES OF THE 3 ";
00375 : PRINT "CONTAINERS!
00380 : INPUT "BEGINNING VOLUME OF GREATEST CONTAINER"; BG
00390 : IF BG < 0 OR BG > CG THEN 380
00400 : INPUT "BEGINNING VOLUME OF MIDDLE CONTAINER"; BM
00410 : IF BM < 0 OR BM > CM THEN 400
00420 : INPUT "BEGINNING VOLUME OF SMALL CONTAINER"; BS
00430 : IF BS < 0 OR BS > CS THEN 420
00440 :
00450 : PRINT
00460 : PRINT "ENTER THE DESIRED END VOLUMES!
00470 : INPUT "END VOLUME OF GREATEST CONTAINER   "; EG
00480 : IF EG < 0 OR EG > CG THEN 470
00490 : INPUT "END VOLUME OF MIDDLE CONTAINER      "; EM
00500 : IF EM < 0 OR EM > CM THEN 490
00510 : INPUT "END VOLUME OF SMALL CONTAINER       "; ES
00520 : IF ES < 0 OR ES > CS THEN 510
00530 : IF EG + EM + ES > BG + BM + BS THEN 450
00540 :
00550 : LET G = BG : LET M = BM : LET S = BS
00560 : PRINT : PRINT : PRINT : PRINT : PRINT G;M;S
00570 :
00580 : LET N = 0 : REM NUMBER OF POURS AT START
00590 :
00600 REM === POURING PROCESS (WITH DISPLAY) ===============
00610 :
00620 : REM --- SELECTION OF CONTAINER TO BE EMPTIED
00630 :
00640 :    IF G = CG THEN 800 : REM POUR GREATEST => SMALL
00650 :    IF M = CM THEN 880 : REM POUR MIDDLE => GREATEST
00660 :    IF S = CS THEN 960 : REM POUR SMALL => MIDDLE
00670 :    IF G = 0  THEN 880 : REM POUR MIDDLE => GREATEST
00680 :    IF M = 0  THEN 960 : REM POUR SMALL => MIDDLE
00690 :    IF S = 0  THEN 800 : REM POUR GREATEST => SMALL
00700 :
```

```
00800 : REM --- POURING FROM GREATEST TO SMALLEST CONTAINER
00810 :
00820 :    LET A = G : LET B = S : LET CB = CS   : REM SET
00825 :                                            REM PARAMETER
00830 :    GOSUB 1120                           : REM POUR
00840 :    LET G = A : LET S = B       : REM SET PARAMETER
00850 :    PRINT G;M;S                         : REM DISPLAY
00860 :    GOTO 1030                    : REM TEST FOR END
00870 :
00880 : REM --- POURING FROM MIDDLE TO GREATEST CONTAINER
00890 :
00900 :    LET A = M : LET B = G : LET CB = CG   : REM SET
00905 :                                            REM PARAMETER
00910 :    GOSUB 1120                           : REM POUR
00920 :    LET M = A : LET G = B       : REM SET PARAMETER
00930 :    PRINT G;M;S                         : REM DISPLAY
00940 :    GOTO 1030                    : REM TEST FOR END
00950 :
00960 : REM --- POURING FROM SMALL TO MIDDLE CONTAINER
00970 :
00980 :    LET A = S : LET B = M : LET CB = CM   : REM SET
00985 :                                            REM PARAMETER
00990 :    GOSUB 1120                           : REM POUR
01000 :    LET S = A : LET M = B       : REM SET PARAMETER
01010 :    PRINT G;M;S                         : REM DISPLAY
01020 :
01030 : REM --- TEST FOR END
01040 :
01050 :    IF NOT (G = EG AND M = EM AND S = ES) THEN 600
01060 :
01070 :    PRINT
01080 :    PRINT "TOTAL OF "; N;" POURS.
01090 :
01100 :    END
01110 :       '
01120 REM +++ SUBROUTINE 'POURING' +++++++++++++++++++++++++
01130 :
01140 : REM THE CONTENTS OF CONTAINER A IS POURED INTO
01150 : REM CONTAINER B. CB IS THE CAPACITY OF CONTAINER B
01160 :
01170 :    IF A+B < CB THEN B = A + B : LET A = 0 : GOTO 1200
01180 :    LET A = A - (CB-B) : LET B = CB
01190 :
01200 :    LET N = N + 1 : REM CONTINUE COUNTING
01210 :
01220 :    RETURN
01230 :
01240 REM END OF POURING +++++++++++++++++++++++++++++++++++
01250 :
```

20. Program the second solution to the Pouring Puzzle with three containers.

21. Rewrite the program for the Pouring Puzzle to include four or more containers.

22. Write a program using the model of the Fifteen Blocks that uses a search process to lead to the equal division of the wine.

23. Here's a problem for you to try to prove using examples. If the capacities of the three containers are relatively prime whole numbers, then pouring back and forth among them can result in every possible gallon amount between 1 and the capacity of the largest container.

****

The crowning puzzle in this chapter on games of patience and concentration involves the most recent brainteaser to hit the market: Rubik's Cube. In no time it took the world of game strategy, mathematics, and computers by storm. Its influence has been equal to that of the Fifteen Blocks which almost caused a mass psychosis at the turn of the century.

Both games are similar in spirit. You recall that Fifteen Blocks involved moving around 15 numbered pieces in a square 4x4 frame. Rubik's Cube involves rearranging the colored sections on a cube measuring 3x3 so that each face of the cube shows one solid color. Both games require the patient player to undo a lot of hard work at an advanced stage of the game in order to finally achieve the winning position. There is no way to "win" either game without going through this frustrating interim process of destroying that almost winning arrangement of blocks or colored cubes.

Example 8.6   RUBIK'S CUBE
The object is to write a program to simulate the Hungarian cube.

The program described below has an advantage over the real cube in that the original arrangement (starting position) of the cube's parts can easily be reconstructed by interrupting the program and beginning over again. Unlike the Fifteen Blocks this isn't a search program, but rather a program meant to aid the player in examininng the cube. To make this function possible, the cube is displayed on the screen in a three-dimensional diagram:

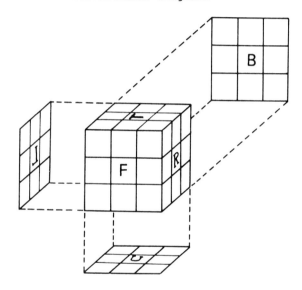

Letters label each face of the cube: F (front); B (back); R (right); L (left); T (top); U (underneath). When the appropriate face of the cube is named, the face is rotated 90 degrees clockwise. Using these operations you can manipulate the cube to get all the combinations.

```
00100 : PRINT CHR$(147)
00110 : PRINT "       RUBIK'S CUBE
00120 : PRINT "       -----------
00130 :
00140 REM SIMULATION OF RUBIKS CUBE
00150 :
```

```
00160 :
00170 REM *** MAIN PROGRAM ********************************
00180 :
00190 : GOSUB 370   : REM DIRECTIONS
00200 :
00210 : GOSUB 700   : REM INITIALIZATION
00220 :
00230 : GOSUB 960   : REM DISPLAY THE CUBE
00240 :
00250 : GOSUB 2000 : REM ENTER MOVE
00260 :
00270 : GOSUB 2260 : REM CALCULATING THE MOVE
00280 :
00290 : GOSUB 4000 : REM COMPLETING THE MOVE
00300 :
00310 : GOTO 250    : REM NEXT MOVE
00320 :
00330 :
00340 REM END OF MAIN PROGRAM ******************************
00350 :
00360 :
00370 REM ### PROCEDURE 'DIRECTIONS' ######################
00380 :
00385 : PRINT : PRINT : PRINT : PRINT
00390 : PRINT "WOULD YOU LIKE TO READ THE DIRECTIONS? (Y/N)
00400 :
00410 : GET A$ : IF A$ =  "N" THEN RETURN
00420 :          IF A$ <> "Y" THEN 410
00430 :
00440 : PRINT CHR$(147);
00450 : PRINT "THIS PROGRAM IS A SIMULATION OF
00460 : PRINT "RUBIK'S CUBE.  IT RUNS ON A
00470 : PRINT "COMMODORE WITH AT LEAST 16 K-BYTE RAM.
00480 :
00490 : PRINT
00500 : PRINT "ENTERING THE MOVE IS REMARKABLY SIMPLE.
00510 : PRINT "ONLY ONE KIND OF TURNING IS POSSIBLE.
00520 : PRINT "ALL SURFACES CAN BE MANIPULATED BY SIMPLY
00530 : PRINT "ENTERING THE FIRST LETTER OF THE SIDE.
00540 : PRINT "THE CUBE WILL MOVE CLOCK-WISE.
00550 : PRINT
00560 : PRINT "TO STOP THE PROGRAM, PRESS 'Q'.
00570 : PRINT
00580 : PRINT "HIT THE SPACE BAR.
00590 :
00600 : GET A$ : IF A$ <> " " THEN 600
00610 :
00620 : RETURN
00630 :
00640 REM END OF DIRECTIONS ##############################
00650 :
00660 :
```

```
00700 REM ### PROCEDURE 'INITIALIZATION' #################
00710 :
00720 : DIM W(6,3,3) : REM FIELDS ON CUBE
00730 :
00740 :    FOR I = 1 TO 6
00750 :      FOR J = 1 TO 3
00760 :        FOR K = 1 TO 3
00770 :          LET W(I,J,K) = I
00780 :        NEXT K
00790 :      NEXT J
00800 :    NEXT I
00810 :
00820 : REM      SURFACE CODES:
00830 : REM      -------------
00840 : REM      FRONT........1
00850 : REM      RIGHT........2
00860 : REM      BACK.........3
00870 : REM      LEFT.........4
00880 : REM      TOP..........5
00890 : REM      UNDERNEATH...6
00900 :
00910 :
00920 : RETURN
00930 :
00940 REM END OF INITIALIZATION ##########################
00950 :
00960 :REM ### PROCEDURE 'DISPLAY OF CUBE' ################
00970 :
00980 : PRINT CHR$(147);
00990 :
01000 : PRINT"
01010 : PRINT"
01020 : PRINT"
01030 : PRINT"
01040 : PRINT"
01050 : PRINT"
01060 : PRINT"
01070 : PRINT"
01080 : PRINT"
01090 : PRINT"
01100 : PRINT"
01110 : PRINT"
01120 : PRINT"
01130 : PRINT"
01140 : PRINT"
01150 : PRINT"
01160 : PRINT"
01170 : PRINT"
01180 : PRINT"
01190 : PRINT"
01200 : PRINT"
01210 : PRINT"
```

301

```
01220 : PRINT"
01230 :
01240 : GOSUB 4000 : REM LABELLING
01250 :
01260 : RETURN
01270 :
01280 REM END OF DISPLAY OF CUBE ##########################
01290 :
01300 :
02000 REM ### PROCEDURE 'ENTER MOVE' #######################
02005 :
02010 : GOSUB 5000 : REM CURSOR CONTROL STRINGS CD$, CL$
02015 :
02020 : PRINT CHR$(19) LEFT$(CD$,10) TAB(20) "YOUR MOVE
02030 : PRINT TAB(20) "(F,R,B,L,T,U):    " LEFT$(CL$,2)
02040 :
02050 : REM --- KEYBOARD QUESTION-AND-ANSWER
02060 :
02070 :    GET A$ : IF A$ = "" THEN 2070
02080 :
02090 :    IF A$ = "F" THEN LET SE = 1 : GOTO 2190
02100 :    IF A$ = "R" THEN LET SE = 2 : GOTO 2190
02110 :    IF A$ = "B" THEN LET SE = 3 : GOTO 2190
02120 :    IF A$ = "L" THEN LET SE = 4 : GOTO 2190
02130 :    IF A$ = "T" THEN LET SE = 5 : GOTO 2190
02140 :    IF A$ = "U" THEN LET SE = 6 : GOTO 2190
02150 :    IF A$ = "Q" THEN PRINT "END." : END : REM EXIT
02160 :
02170 :    GOTO 2070
02180 :
02190 :    PRINT A$
02200 :
02210 :    RETURN
02220 :
02230 REM ### END OF ENTER MOVE ###########################
02240 :
02250 :
02260 REM ### PROCEDURE 'CALCULATE MOVE' ##################
02270 :
02280 : REM --- TURN SURFACE
02290 :
02300 :    LET Z         = W(SE,3,1)
02310 :    LET W(SE,3,1) = W(SE,1,1)
02320 :    LET W(SE,1,1) = W(SE,1,3)
02330 :    LET W(SE,1,3) = W(SE,3,3)
02340 :    LET W(SE,3,3) = Z
02350 :
02360 :    LET Z         = W(SE,3,2)
02370 :    LET W(SE,3,2) = W(SE,2,1)
02380 :    LET W(SE,2,1) = W(SE,1,2)
02390 :    LET W(SE,1,2) = W(SE,2,3)
02400 :    LET W(SE,2,3) = Z
```

```
02410 :
02420 : REM --- TURN EDGE
02430 :
02440 :    ON SE GOSUB 2480,2700,2920,3140,3360,3580
02450 :
02460 :    RETURN
02470 :
02480 : REM --- FRONT (1)
02490 :
02500 :    LET Z         = W(5,1,1)
02510 :    LET W(5,1,1) = W(4,1,3)
02520 :    LET W(4,1,3) = W(6,1,1)
02530 :    LET W(6,1,1) = W(2,3,1)
02540 :    LET W(2,3,1) = Z
02550 :
02560 :    LET Z         = W(5,1,2)
02570 :    LET W(5,1,2) = W(4,2,3)
02580 :    LET W(4,2,3) = W(6,1,2)
02590 :    LET W(6,1,2) = W(2,2,1)
02600 :    LET W(2,2,1) = Z
02610 :
02620 :    LET Z         = W(5,1,3)
02630 :    LET W(5,1,3) = W(4,3,3)
02640 :    LET W(4,3,3) = W(6,1,3)
02650 :    LET W(6,1,3) = W(2,1,1)
02660 :    LET W(2,1,1) = Z
02670 :
02680 :    RETURN
02690 :
02700 : REM --- RIGHT (2)
02710 :
02720 :    LET Z         = W(5,1,3)
02730 :    LET W(5,1,3) = W(1,1,3)
02740 :    LET W(1,1,3) = W(6,3,1)
02750 :    LET W(6,3,1) = W(3,3,1)
02760 :    LET W(3,3,1) = Z
02770 :
02780 :    LET Z         = W(5,2,3)
02790 :    LET W(5,2,3) = W(1,2,3)
02800 :    LET W(1,2,3) = W(6,2,1)
02810 :    LET W(6,2,1) = W(3,2,1)
02820 :    LET W(3,2,1) = Z
02830 :
02840 :    LET Z         = W(5,3,3)
02850 :    LET W(5,3,3) = W(1,3,3)
02860 :    LET W(1,3,3) = W(6,1,1)
02870 :    LET W(6,1,1) = W(3,1,1)
02880 :    LET W(3,1,1) = Z
02890 :
02900 :    RETURN
02910 :
02920 : REM --- BACK (3)
```

303

```
02930 :
02940 :    LET Z        = W(5,3,3)
02950 :    LET W(5,3,3) = W(2,1,3)
02960 :    LET W(2,1,3) = W(6,3,3)
02970 :    LET W(6,3,3) = W(4,3,1)
02980 :    LET W(4,3,1) = Z
02990 :
03000 :    LET Z        = W(5,3,2)
03010 :    LET W(5,3,2) = W(2,2,3)
03020 :    LET W(2,2,3) = W(6,3,2)
03030 :    LET W(6,3,2) = W(4,2,1)
03040 :    LET W(4,2,1) = Z
03050 :
03060 :    LET Z        = W(5,3,1)
03070 :    LET W(5,3,1) = W(2,3,3)
03080 :    LET W(2,3,3) = W(6,3,1)
03090 :    LET W(6,3,1) = W(4,1,1)
03100 :    LET W(4,1,1) = Z
03110 :
03120 :    RETURN
03130 :
03140 : REM --- LEFT (4)
03150 :
03160 :    LET Z        = W(5,3,1)
03170 :    LET W(5,3,1) = W(1,3,1)
03180 :    LET W(1,3,1) = W(6,1,3)
03190 :    LET W(6,1,3) = W(3,1,3)
03200 :    LET W(3,1,3) = Z
03210 :
03220 :    LET Z        = W(5,2,1)
03230 :    LET W(5,2,1) = W(1,2,1)
03240 :    LET W(1,2,1) = W(6,2,3)
03250 :    LET W(6,2,3) = W(3,2,3)
03260 :    LET W(3,2,3) = Z
03270 :
03280 :    LET Z        = W(5,1,1)
03290 :    LET W(5,1,1) = W(1,1,1)
03300 :    LET W(1,1,1) = W(6,3,3)
03310 :    LET W(6,3,3) = W(3,3,3)
03320 :    LET W(3,3,3) = Z
03330 :
03340 :    RETURN
03350 :
03360 : REM --- TOP (5)
03370 :
03380 :    LET Z        = W(1,3,3)
03390 :    LET W(1,3,3) = W(2,3,3)
03400 :    LET W(2,3,3) = W(3,3,3)
03410 :    LET W(3,3,3) = W(4,3,3)
03420 :    LET W(4,3,3) = Z
03430 :
03440 :    LET Z        = W(1,3,2)
```

```
03450 :    LET W(1,3,2) = W(2,3,2)
03460 :    LET W(2,3,2) = W(3,3,2)
03470 :    LET W(3,3,2) = W(4,3,2)
03480 :    LET W(4,3,2) = Z
03490 :
03500 :    LET Z        = W(1,3,1)
03510 :    LET W(1,3,1) = W(2,3,1)
03520 :    LET W(2,3,1) = W(3,3,1)
03530 :    LET W(3,3,1) = W(4,3,1)
03540 :    LET W(4,3,1) = Z
03550 :
03560 :    RETURN
03570 :
03580 : REM --- UNDERNEATH (6)
03590 :
03600 :    LET Z        = W(1,1,1)
03610 :    LET W(1,1,1) = W(4,1,1)
03620 :    LET W(4,1,1) = W(3,1,1)
03630 :    LET W(3,1,1) = W(2,1,1)
03640 :    LET W(2,1,1) = Z
03650 :
03660 :    LET Z        = W(1,1,2)
03670 :    LET W(1,1,2) = W(4,1,2)
03680 :    LET W(4,1,2) = W(3,1,2)
03690 :    LET W(3,1,2) = W(2,1,2)
03700 :    LET W(2,1,2) = Z
03710 :
03720 :    LET Z        = W(1,1,3)
03730 :    LET W(1,1,3) = W(4,1,3)
03740 :    LET W(4,1,3) = W(3,1,3)
03750 :    LET W(3,1,3) = W(2,1,3)
03760 :    LET W(2,1,3) = Z
03770 :
03780 :    RETURN
03790 :
03800 :
03810 REM ### END OF CALCULATING MOVE ####################
03820 :
03830 :
04000 REM ### PROCEDURE 'COMPLETE MOVE' ##################
04010 :
04020 : REM --- SURFACE BACK (3)
04030 :
04040 :    PRINT CHR$(19) CHR$(17) TAB(17);
04050 :    FOR I = 3 TO 1 STEP -1
04060 :      FOR J = 3 TO 1 STEP -1
04070 :        PRINT RIGHT$(STR$(W(3,I,J)),1) CHR$(29);
04080 :      NEXT J
04090 :      PRINT LEFT$(CL$,6) LEFT$(CD$,2);
04100 :    NEXT I
04110 :
04120 : REM --- SURFACE LEFT (4)
```

305

```
04130 :
04140 :    PRINT CHR$(19) LEFT$(CD$,12) CHR$(29);
04150 :    FOR I = 3 TO 1 STEP -1
04160 :      FOR J = 3 TO 1 STEP -1
04170 :        PRINT RIGHT$(STR$(W(4,I,J)),1) CHR$(145);
04180 :      NEXT J
04190 :      PRINT LEFT$(CL$,3) LEFT$(CD$,5);
04200 :    NEXT I
04210 :
04220 ; REM --- SURFACE TOP (5)
04230 :
04240 :    PRINT CHR$(19) LEFT$(CD$,7)
04250 :    FOR I = 3 TO 1 STEP -1
04260 :      PRINT TAB(6+I);
04270 :      FOR J = 1 TO 3
04280 :        PRINT RIGHT$(STR$(W(5,I,J)),1) CHR$(29);
04290 :      NEXT J
04300 :      PRINT
04310 :    NEXT I
04320 :
04330 : REM --- SURFACE FRONT (1)
04340 :
04350 :    PRINT CHR$(19) LEFT$(CD$,11)
04360 :    FOR I = 3 TO 1 STEP -1
04370 :      PRINT TAB(6);
04380 :      FOR J = 1 TO 3
04390 :        PRINT RIGHT$(STR$(W(1,I,J)),1) CHR$(29);
04400 :      NEXT J
04410 :      PRINT CHR$(17)
04420 :    NEXT I
04430 :
04440 : REM --- SURFACE RIGHT (2)
04450 :
04460 :    PRINT CHR$(19) LEFT$(CD$,12) TAB(12);
04470 :    FOR I = 3 TO 1 STEP -1
04480 :      FOR J = 1 TO 3
04490 :        PRINT RIGHT$(STR$(W(2,I,J)),1) CHR$(145);
04500 :      NEXT J
04510 :      PRINT LEFT$(CL$,3) LEFT$(CD$,5);
04520 :    NEXT I
04530 :
04540 : REM --- SURFACE UNDERNEATH (6)
04550 :
04560 :    PRINT CHR$(19) LEFT$(CD$,18)
04570 :    FOR I = 3 TO 1 STEP -1
04580 :      PRINT TAB(6+I);
04590 :      FOR J = 3 TO 1 STEP -1
04600 :        PRINT RIGHT$(STR$(W(6,I,J)),1) CHR$(29);
04610 :      NEXT J
04620 :      PRINT
04630 :    NEXT I
04640 :
```

```
04650 :    RETURN
04660 :
04670 REM END OF PROCEDURE 'COMPLETE MOVE' ################
04680 :
05000 REM SUBROUTINE 'CURSOR CONTROL STRINGS'===============
05010 :
05020 : LET CD$ = CHR$(17) : LET CL$ = CHR$(157)
05030 : FOR I = 1 TO 5
05040 :    LET CD$ = CD$ + CD$ : LET CL$ = CL$ + CL$
05050 : NEXT I
05060 :
05070 : RETURN
05080 :
05090 REM END OF SUBROUTINE 'CURSOR CONTROL STRINGS'========
05100 :
```

****

BRAINTEASERS

24. THE TOWER   At the 1899 World Expo-
sition in Paris the mathematician E.
Lucas exhibited some of his mathematical games. Among them was "The Tower."
He told an imaginative tale about the
origin of the game in the far east
(which is why it is also called Tower
of Hanoi).

     Eight round discs are piled on one
of three plates. The largest is on the
bottom, the smallest on top. The object of the game is to rebuild the
tower on another plate by removing the
discs one by one and never letting a
larger disc be placed on top of a smaller one. You can only
place a disc on an empty plate or on a larger disc. Write a
program following these rules to make this possible on the
screen.

25. HEADSCRATCHER   Zeroes (0) and ones (1) are distributed
randomly on a square measuring 3 x 3. A move consists of
stating one of the 9 positions in the field of play. This
will have various results:

307

a) If one of the four corners is stated, then in addition to the corner itself, both of the neighboring corners as well as the middle of the square change. "Change" means to turn 0 into 1 and vice-versa.

b) If one of the four middle squares on a side is named, then both neighboring corners change as well.

c) If the center square is named, then all four middle side squares change too.

Write a program to make it possible to play this on the screen. The goal is to reach a winning position that looks like this:

```
1   1   1
1   0   1
1   1   1
```

26. JOSEPH'S GAME   History recounts a brutal incident from the suppression of the Jewish uprising by the Romans (c. 70 A.D.). Forty-one rebels who had found refuge in a cave knew they were in imminent danger of being captured and put to death by their enemies. In order to escape this end, they devised a system of suicide. They formed a circle in which every 9th person should die -- except the last two who were charged with killing each other. The historian Flavius Josephus, who was among the unfortunate band, took his place in the circle so that he and a weak comrade (whom he easily defeated) were the two survivors. How did he do it?

The problem can be generalized like this: N people or objects form a circle; every Kth one is eliminated and the circle becomes smaller. The object is to find the order of elimination. In later times, especially in the Middle Ages, this was a popular "counting-out" game with many variations.

Is it possible to arrange the people so that the counting out eliminates a subset, such as only men or only women?

CHAPTER IX

MIXED BRAINTEASERS

In this concluding chapter you'll find a selection of
activities of varying degrees of difficulty. These should
be fairly easy to solve with the computer's help. By now
you are probably having enough fun programming anyway so
that you don't have to be reassured that these brainteasers
are easy.

1. In an office of a well-known publisher in Cambridge,
Massachusetts there is a personal computer. Its serial number has 5 digits, all different, and is divisible by 9. The
first digit is even and is the product of the third and
fourth digits. The sum of the first two digits is 15; the
third digit is the difference of the first two. What is the
serial number?

2. The telephone directory of a major city has fewer than
1000 pages but exactly 999,991 names listed. Every page
lists the same number of names. How many pages does the
book have?

3. Otto is a dependable boy when it comes to mathematics.
Rest assured that when he says there is only one other solution to a problem, he has thought of every angle. He wants
you to come up with a 7-digit number, which, when multiplied
by a certain whole number, results in its inversion. His

solution is as follows: 1099989 * 9 = 9899901.   What is
Otto's other number?

4. Here are some mathematical statements:

$$89 = 8^1 + 9^2$$
$$135 = 1^1 + 3^2 + 5^3$$
$$2427 = 2^1 + 4^2 + 2^3 + 7^4$$

Are there other numbers with this characteristic?   Let the
computer do the searching.

5. Take a look at these numbers:

$$81 = 9^2 = (8 + 1)^2$$
$$512 = 8^3 = (5 + 1 + 2)^3$$

These are examples of numbers with N digits that are also
equal to the Nth power of their digit total.   Find examples
for N = 4 and demonstrate that there is no solution when N
= 5.

6. There are six numbers between 1 and 10 that are the solutions to the following equations:

$$A + B + C = D + E + F$$
$$A^2 + B^2 + C^2 = D^2 + E^2 + F^2$$

Five of the numbers are "pairwise different".   Two are
equal.

7. The number 473,684,210,526,315,789 has the remarkable
property of being doubled by moving the 9 from the one's
place to the far left.   Can you find any more examples of
such numbers?

8. Look at this equation:

$$60 * 84 = 35 * (60 + 84)$$

310

This tells us that the sum of 60 and 84 divides its product. Is this the only pair of numbers having this property? Let the computer help you find the answer.

9. Otto is looking for the smallest number with the following property: when multiplied by the number in its own one's place, the result is the same as when the number in its own one's place is moved to the far left.

10. Here is another remarkable number: if you replace the digits in 148 using a cyclical pattern you come up with 481 and 814. Their differences are the same (333). Find more examples of such numbers.

11. Fifteen people are talking about a number. Two of them are lying and the rest are telling the truth. Here is their conversation:

     1. "The number is a multiple of 2."
     2. "The number is a multiple of 3."
     3. "The number is a multiple of 4."
     4. "The number is a multiple of 5."
     5. "The number is a multiple of 6."
     6. "The number is a multiple of 7."
     7. "The number is a multiple of 8."
     8. "The number is a multiple of 9."
     9. "The number is a multiple of 10."
    10. "The number is smaller than 1000."
    11. "The number is smaller than 750."
    12. "The number is smaller than 550."
    13. "The number is smaller than 500."
    14. "The number is greater than 400."
    15. "The number is greater than 450."

What is the number?

12. Four players agree that anybody who loses must pay the others the amount that they already possess. After four rounds each player has already lost once. At the end the

players total up their capital and discover that they all have the same amount of money, namely $160.00. How much did each have before starting the game?

13. It's New Year's Day 1984. Mr. Smith is on the train and across from him sits Mr. Brown. During the course of their conversation they get to talking about age. Smith says he is as old as the sum of the digits of his birth year. After thinking a minute Brown wishes him Happy Birthday. How old is Smith?

14. Six people sit down to dinner. Jill sits to the left of Jack, Alexandra does not sit beside Nicholas, nor to the right of Louise, but rather across from Andrew. Figure out the order of these people at the round table.

15. Put the numbers from 1 through 7 in the diagram so that every row, column, and diagonal adds up to the same total.

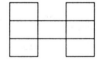

16. Put the numbers from 1 through 9 into the diagram so that they add up to the same total in each triangle.

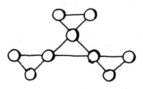

17. The following grouping of numbers is a so-called anti-magic square:

$$
\begin{array}{ccc}
1 & 2 & 3 \\
8 & 9 & 4 \\
7 & 6 & 5
\end{array}
$$

No two rows, columns, or diagonals have the the same total.

Have the computer generate more of these anti-magic squares.
Try some of a higher magnitude.

18. In athletic contests the overall rank of a team is sometimes calculated by adding the team rank of the individual
team players. If they have the rank, let's say, of 1, 3,
and 6, then the team gets the rank of 1 + 3 + 6 = 10. This
means the team has achieved first place because the members
of another team could, at best, have the ranks of 2, 4, 5
which equal a worse team rank of 11. There cannot be two
first places, either for a single player or for a team.
    The rank of 10, moreover, is maximal in the sense that
the next highest rank of 11 does not always result in the
first team rank -- as the example shows. The question now
is: when the team's strength is M ∈ {1,2,3...}, how large
is the maximum rank p_M for the first team place? Hint: you
have to find $p_1 = 1$, $p_2 = 4$, and, as in the example, $p_3 = 10$).

19. The following is known about function f:

$$f(0,y) = y + 1$$
$$f(x+1,0) = f(x,1)$$
$$f(x+1,y+1) = f(x,f(x+1,y))$$

What is f(4,1981)?

20. Figure out how it is possible for the sheep standing in
the first position to get into the second position:

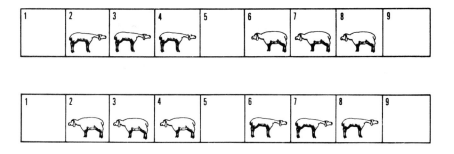

You are allowed to move a sheep to an empty square and jump over a sheep into an empty square. You may not go backward or jump over two sheep. Write a program for this game.

21. Here is a diagram showing 8 checkers.

You have five moves to rearrange them so that the four white ones and the four black ones are grouped together. The catch is that you are allowed to move only two checkers per move; furthermore, these two checkers can do nothing more than exchange places with each other.

22. The object of this game is to exchange the positions on the board of the white and the black knights in as few moves as possible. Devise a search program to do this. (Hint: 16 moves are enough.)

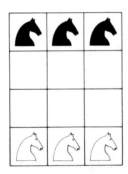

23. The Oxford mathematician Charles Lutwidge Dodgson (1832-1898) (better known under his pseudonym of Lewis Carroll as the author of <u>Alice in Wonderland</u>) wrote in his journal that he had been unsuccessful in finding three right triangles having the same area and whose side lengths were whole numbers. Try finding a solution with the computer.

24. Monica plans to invite 15 friends to dinner and eat with 3 of them each of 35 days. She wants to arrange it so that

two of her friends never meet each other at her place more than once. Can she balance such a complicated schedule? Ask the computer.

25. A convention is being held in a certain town. Delegations from seven countries (each consisting of three old gentlemen) meet at the railroad station. Seven limousines pick them up at the station. Three delegates, each from a different country, are to ride in each car. Your object is to seat them so that the closest contact possible among all delegations is achieved. If you were to divide the delegations up according to these combinations:

   1,2,3; 2,3,4; 3,4,5; 4,5,6; 5,6,7; 6,7,1; 7,1,2

then delegates from country 1 would come into contact with those from countries 2,3,6,7, but not with anyone from countries 4 and 5. How would you have to arrange the trip to achieve these last combinations as well?

26. Seven friends are members of the same poker club. Every evening (even Sundays) three of them get together. They want to keep the bond of friendship as close as possible and therefore have to arrange the combinations so that each player sits down at the table with every other player once a week.

27. Every day of the week a teacher at a boarding school takes a walk with 15 of the pupils. The children walk in an orderly fashion in groups of three. Can the teacher arrange the groups so that no two children, who have once walked together in a trio, ever end up in the same group again that week?

28. Nine dangerous prisoners are only allowed to walk around the courtyard of the penitentiary if they are chained together. On the week days they are chained together in groups of three. Within one week no two prisoners may be chained together more than once. Have the computer find all the combinations possible.

315

316

THE PROGRAMMING LANGUAGE BASIC

Here are the most important commands in BASIC (Commodore version):

ABS assigns a number its absolute value. Example: PRINT ABS(3) results in 3, PRINT ABS(-3) also results in 3.

AND combines two logical expressions so as to give the result of "true" if both expressions are "true."

ASC assigns an ASCII number to a character. Example: PRINT ASC("A") results in 65.

CHR$ is the inverse function of ASC. Every integer between 0 and 255 is assigned its own ASCII character. Example: the command PRINT CHR$(65) produces the letter A.

CLR erases all variables (i.e., assigns to numerical variables the value of 0 and to character sequence variables the value of an empty sequence. It also flushes all dimensionalization of tables and subroutine addresses.

CONT continues the program run after hitting the stop button or after carrying out the commands STOP or END.

DATA/READ/RESTORE are commands that store and read data in a program.

DEF FN defines a function. Example: DEF FN Y(X) = X * X. This function may be invoked with PRINT FN Y(5), so (5 * 5 =) 25 is printed.

DIM sets the dimensions of a table (for preparing memory storage space).

END ends the program run and lets the computer return to

317

direct dialogue.

FOR...NEXT (STEP) programs the counting loop (counted repetition).

GET reads in a single character from the keyboard. The computer does not wait for the key to be hit (in contrast to INPUT).

GOSUB....RETURN jumps into a subroutine and back to the main program.

GOTO is the primary move command.

IF....THEN is the decision command (conditional command).

INPUT reads data from the keyboard.

INT rounds a decimal down to the next smallest whole number. Example: PRINT INT(-3.14) results in -4.

LEFT$ retrieves characters at the left of a character sequence.   Example: PRINT LEFT$("HAYSTACK",3) results in HAY. The numerical argument tells how many characters to extract.

LEN returns the number of characters in a character sequence.

LET is the command that assigns value.

LIST lists a whole program, individual lines, or regions.

MID$ retrieves certain characters or portions of a sequence from a longer character sequence.   Example: PRINT MID$("HAYSTACK",3,3) results in YST.

NEW flushes the entire program.

NOT gives a logical expression its negation.

ON...GOTO / ON...GOSUB is the conditional jump.

OR connects two logical expressions so that the result is "true" when at least one of the expressions is "true".

PEEK lets the user examine a memory cell.

POKE writes data into a memory cell.

PRINT is the command for display on the screen.

REM signals a remark in the program. The computer ignores this line.

RIGHT$ retrieves characters from the right end of a character sequence (string).

RND generates (pseudo) random numbers between 0 and 1.

RUN starts a program.

TAB sets the tab for the display.

The following POKE codes may be useful (codes above 64 are for graphics symbols):

CODE(s)	CHARACTER(s)	DEFINITION
20	T	capital t
30	↑	up arrow
32		blank space
35	#	number sign
36	$	dollar sign
37	%	percent sign
42	*	asterick
43	+	plus sign
46	.	period
48-57	0-9	digits 0 thru 9
60	<	less than sign
61	=	equals sign
62	>	greater than sign
63	?	question mark
64	—	(heavy dash)
65	♠	(spade)
66	|	(vertical line)
81	●	(heavy ball)
83	♥	(heart)
86	X	(heavy X)
87	O	(ball)
88	♣	(club)
90	♦	(diamond)
93	|	(vertical line)
98	■	(block)
102	▒	(checkerboard)
103	|	(vertical line right)
111	▬	(heavy dash bottom)
121	▪	(heavier dash bottom)
122	⌟	(bottom right corner)
160	█	(inverse blank space)
179	3	(inverse 3)
193	♠	(inverse spade)

Attention all game players! Whatever your pleasure—games of strategy, skill, choice, chance, logic—*BASIC Game Plans* can teach you how to program your computer to play, with you or against you. Anything's possible, from simple tic-tac-toe to the most challenging of moving objects—even games that haven't been invented yet!

Whether you're a hyper hacker or a nervous novice, this book has programs for you. Programming the computer itself becomes the game!

ISBN 0-8176-3366-9
ISBN 3-7643-3366-9